Probability and Mathematical Statistics (Continued)

PURI, VILAPLANA, and WERTZ • New Perspectives in Theoretical and Applied Statistics

RANDLES and WOLFE • Introduction to the Theory of Nonparametric Statistics

RAO • Linear Statistical Inference and Its Applications, *Second Edition*

RAO • Real and Stochastic Analysis

RAO and SEDRANSK • W.G. Cochran's Impact on Statistics

RAO • Asymptotic Theory of Statistical Inference

ROHATGI • An Introduction to Probability Theory and Mathematical Statistics

ROHATGI • Statistical Inference

ROSS • Stochastic Processes

RUBINSTEIN • Simulation and The Monte Carlo Method

SCHEFFE • The Analysis of Variance

SEBER • Linear Regression Analysis

SEBER • Multivariate Observations

SEN • Sequential Nonparametrics: Invariance Principles and Statistical Inference

SERFLING • Approximation Theorems of Mathematical Statistics

SHORACK and WELLNER • Empirical Processes with Applications to Statistics

TJUR • Probability Based on Radon Measures

Applied Probability and Statistics

ABRAHAM and LEDOLTER • Statistical Methods for Forecasting

AGRESTI • Analysis of Ordinal Categorical Data

AICKIN • Linear Statistical Analysis of Discrete Data

ANDERSON, AUQUIER, HAUCK, OAKES, VANDAELE, and WEISBERG • Statistical Methods for Comparative Studies

ARTHANARI and DODGE • Mathematical Programming in Statistics

ASMUSSEN • Applied Probability and Queues

BAILEY • The Elements of Stochastic Processes with Applications to the Natural Sciences

BAILEY • Mathematics, Statistics and Systems for Health

BARNETT • Interpreting Multivariate Data

BARNETT and LEWIS • Outliers in Statistical Data, *Second Edition*

BARTHOLOMEW • Stochastic Models for Social Processes, *Third Edition*

BARTHOLOMEW and FORBES • Statistical Techniques for Manpower Planning

BECK and ARNOLD • Parameter ⸱⸱⸱⸱⸱⸱⸱⸱ and Science

BELSLEY, KUH, and WELSCH ⸱⸱⸱⸱⸱⸱⸱⸱⸱g Influential Data and Sources of C

BHAT • Elements of Applied Stoc

BLOOMFIELD • Fourier Analysi

BOX • R. A. Fisher, The Life of a

BOX and DRAPER • Empirical

BOX and DRAPER • Evolutionary ⸱⸱⸱⸱⸱⸱ for Process Improvement

BOX, HUNTER, and HUNTER • Statistics for Experimenters: An Introduction to Design, Data Analysis, and Model Building

BROWN and HOLLANDER • Statistics: A Biomedical Introduction

BUNKE and BUNKE • Statistical Inference in Linear Models, Volume I

CHAMBERS • Computational Methods for Data Analysis

CHATTERJEE and PRICE • Regression Analysis by Example

CHOW • Econometric Analysis by Control Methods

CLARKE and DISNEY • Probability and Random Processes: A First Course with Applications, *Second Edition*

COCHRAN • Sampling Techniques, *Third Edition*

COCHRAN and COX • Experimental Designs, *Second Edition*

CONOVER • Practical Nonparametric Statistics, *Second Edition*

CONOVER and IMAN • Introduction to Modern Business Statistics

CORNELL • Experiments with Mixtures: Designs, Models and The Analysis of Mixture Data

D1293657

Applied Probability and Statistics (Continued)

COX • Planning of Experiments

COX • A Handbook of Introductory Statistical Methods

DANIEL • Biostatistics: A Foundation for Analysis in the Health Sciences, *Fourth Edition*

DANIEL • Applications of Statistics to Industrial Experimentation

DANIEL and WOOD • Fitting Equations to Data: Computer Analysis of Multifactor Data, *Second Edition*

DAVID • Order Statistics, *Second Edition*

DAVISON • Multidimensional Scaling

DEGROOT, FIENBERG and KADANE • Statistics and the Law

DEMING • Sample Design in Business Research

DILLON and GOLDSTEIN • Multivariate Analysis: Methods and Applications

DODGE • Analysis of Experiments with Missing Data

DODGE and ROMIG • Sampling Inspection Tables, *Second Edition*

DOWDY and WEARDEN • Statistics for Research

DRAPER and SMITH • Applied Regression Analysis, *Second Edition*

DUNN • Basic Statistics: A Primer for the Biomedical Sciences, *Second Edition*

DUNN and CLARK • Applied Statistics: Analysis of Variance and Regression

ELANDT-JOHNSON and JOHNSON • Survival Models and Data Analysis

FLEISS • Statistical Methods for Rates and Proportions, *Second Edition*

FLEISS • The Design and Analysis of Clinical Experiments

FOX • Linear Statistical Models and Related Methods

FRANKEN, KÖNIG, ARNDT, and SCHMIDT • Queues and Point Processes

GALLANT • Nonlinear Statistical Models

GIBBONS, OLKIN, and SOBEL • Selecting and Ordering Populations: A New Statistical Methodology

GNANADESIKAN • Methods for Statistical Data Analysis of Multivariate Observations

GREENBERG and WEBSTER • Advanced Econometrics: A Bridge to the Literature

GROSS and HARRIS • Fundamentals of Queueing Theory, *Second Edition*

GUPTA and PANCHAPAKESAN • Multiple Decision Procedures: Theory and Methodology of Selecting and Ranking Populations

GUTTMAN, WILKS, and HUNTER • Introductory Engineering Statistics, *Third Edition*

HAHN and SHAPIRO • Statistical Models in Engineering

HALD • Statistical Tables and Formulas

HALD • Statistical Theory with Engineering Applications

HAND • Discrimination and Classification

HOAGLIN, MOSTELLER and TUKEY • Exploring Data Tables, Trends and Shapes

HOAGLIN, MOSTELLER, and TUKEY • Understanding Robust and Exploratory Data Analysis

HOEL • Elementary Statistics, *Fourth Edition*

HOEL and JESSEN • Basic Statistics for Business and Economics, *Third Edition*

HOGG and KLUGMAN • Loss Distributions

HOLLANDER and WOLFE • Nonparametric Statistical Methods

IMAN and CONOVER • Modern Business Statistics

JAGERS • Branching Processes with Biological Applications

JESSEN • Statistical Survey Techniques

JOHNSON • Multivariate Statistical Simulation

JOHNSON and KOTZ • Distributions in Statistics

                                    Discrete Distributions

                                    Continuous Univariate Distributions—1

                                    Continuous Univariate Distributions—2

                                    Continuous Multivariate Distributions

(*continued on back*)

Small Area Statistics

Small Area Statistics

AN INTERNATIONAL SYMPOSIUM

Edited by

R. PLATEK

Census and Household Survey Methods Division
Statistics Canada
Ottawa, Ontario, Canada

J. N. K. RAO

Department of Mathematics and Statistics
Carleton University
Ottawa, Ontario, Canada

C. E. SÄRNDAL

Département de mathématiques et de statistique
Université de Montréal
Montréal, Québec, Canada

M. P. SINGH

Methodology Branch
Statistics Canada
Ottawa, Ontario, Canada

JOHN WILEY & SONS
New York • Chichester • Brisbane • Toronto • Singapore

Library of Congress Cataloging in Publication Data:
Small area statistics.

(Wiley series in probability and mathematical
statistics. Applied probability and statistics)
    Includes index.
    1. Social sciences--Statistical methods--
Congresses. I. Platek, Richard. II. Series.

HA29.S5775   1987   300'.1'5195   86-13221
ISBN 0-471-84456-X

Printed in the United States of America

10 9 8 7 6 5 4 3 2 1

Contributors

K. G. Basavarajappa, Demography Division, Statistics Canada, Ottawa, Ontario, Canada

Gordon J. Brackstone, Methodology Branch, Statistics Canada, Ottawa, Ontario, Canada

Claes-M. Cassel, I/UI, Statistics Sweden, Stockholm, Sweden

Fred R. Cronkhite, Office of Employment and Unemployment Statistics, U.S. Department of Labor, Washington, D.C.

Estela B. Dagum, Time Series Research and Analysis Division, Statistics Canada, Ottawa, Ontario, Canada

Tore Dalenius, Division of Applied Mathematics, Brown University, Providence, Rhode Island

Arthur P. Dempster, Department of Statistics, Science Centre, Cambridge, Massachusetts

Gregg J. Diffendal, Statistical Research Division, Department of Commerce, Bureau of the Census, Washington, D.C.

Eugene P. Ericksen, Department of Sociology, Temple University, Philadelphia, Pennsylvania

Robert E. Fay, Statistical Methods Division, Department of Commerce, Bureau of the Census, Washington, D.C.

Gregory A. Feeney, Sampling (Social Statistics) Section, Australian Bureau of Statistics, Belconnen, Australia

Wayne A. Fuller, Department of Statistics, Iowa State University, Ames, Iowa

Rachel M. Harter, Department of Statistics, Iowa State University, Ames, Iowa

v

MICHAEL A. HIDIROGLOU, Business Survey Methods Division, Statistics Canada, Ottawa, Ontario, Canada

CARY T. ISAKI, Statistical Research Division, Department of Commerce, Bureau of the Census, Washington, D.C.

JOSEPH B. KADANE, Department of Statistics, Carnegie-Mellon University, Pittsburgh, Pennsylvania

GRAHAM KALTON, Survey Research Center, Institute of Social Research, University of Michigan, Ann Arbor, Michigan

LESLIE KISH, Institute of Social Research, University of Michigan, Ann Arbor, Michigan

KARL-ERIK KRISTIANSSON, I/UI, Statistics Sweden, Stockholm, Sweden

SIXTEN LUNDSTRÖM, P/STM, Statistics Sweden, Orebro, Sweden

PETER MCCULLAGH, Department of Statistics, University of Chicago, Chicago, Illinois

MARIETTA A. MORRY, Time Series Research and Analysis Division, Statistics Canada, Ottawa, Ontario, Canada

GÖRAN RÅBÄCK, I/UI, Statistics Sweden, Stockholm, Sweden

T. E. RAGHUNATHAN, Department of Statistics, Science Centre, Cambridge, Massachusetts

LINDA KAY SCHULTZ, Statistical Research Division, Department of Commerce, Bureau of the Census, Washington, D.C.

PHILLIP J. SMITH, Statistical Research Division, Department of Commerce, Bureau of the Census, Washington, D.C.

THOMAS W. F. STROUD, Department of Mathematics and Statistics, Queen's University, Kingston, Ontario, Canada

RAVI B. P. VERMA, Demography Division, Statistics Canada, Ottawa, Ontario, Canada

STAFFAN WAHLSTRÖM, I/UI, Statistics Sweden, Stockholm, Sweden

JAMES V. ZIDEK, Department of Mathematics and Statistics, University of British Columbia, Vancouver, B.C., Canada

Preface

In recent years, the demand for small area statistics has greatly increased. This is due, among other things, to their growing use in formulating policies and programs, in the dispensation of government funds, and in regional planning. Legislative acts by national governments have also increasingly created a need for small area statistics, and this trend will most likely continue. As a consequence, the production of reliable small area statistics has emerged as a pressing and frequently difficult and costly problem.

Several agencies of federal and provincial governments, including Statistics Canada, have introduced vigorous programs to meet this new demand, with a view toward producing efficient and high quality statistics. Furthermore, significant research on both the theoretical and the practical aspects of small area estimation is conducted at various universities and in some national statistical bureaus.

As a result of joint initiative of Statistics Canada, The Laboratory for Research in Statistics and Probability of Carleton University, and the Département de mathématiques et de statistique of the Université de Montréal, an International Symposium on Small Area Statistics was organized. The symposium was held in May 1985, at Canada's Capital Congress Centre in Ottawa.

Roughly 300 registered participants benefited from the presentations of invited and contributed papers by prominent statisticians and survey researchers representing universities and statistical agencies throughout the world. A panel discussion was also held.

The present volume contains all the invited papers presented at the symposium. A separate volume containing the contributed papers will be issued elsewhere. All papers were refereed and were also reviewed by the co-editors.

Presentation in each session was followed by a discussion period.

Although interesting views and comments emerged, for practical reasons, these discussions are not included in the present volume.

The reader will find, in the papers that follow, various levels of practical and theoretical undertones. We believe it to be of value at the present stage of development to provide a cross section of the various challenges that face small area estimation.

The content of the volume is organized into five parts:

Part 1: Policy issues

Part 2: Population estimation for small areas

Part 3: Theoretical developments

Part 4: Organizational experiences with small area estimation techniques

Part 5: Panel discussion

In Part I, "Policy Issues," Brackstone addresses issues arising in the development and provision of geographically integrated small area data to meet an increasing volume of consumer demand. From the perspective of a national statistical agency, the paper illustrates various issues by tracing the conception and early development of the Small Area Data Program in Statistics Canada. This program embraces a wide variety of activities concerned with data development, data organization and dissemination, as well as the provision of required infrastructure for these activities.

Part 2, "Population Estimation for Small Areas," consists of papers by Ericksen and Kadane, Verma and Basavarajappa, and McCullagh and Zidek.

Ericksen and Kadane are concerned in their paper with local estimation of the population undercount in the 1980 U.S. Census. The general method is to obtain sample estimates of the undercount for a set of local areas and to use regression to relate these to a set of predictor variables. For local areas with sample data, a final estimate is computed as a weighted average of sample and regression estimates. For the remaining areas, the regression equation is applied. The authors used this method when appearing as plaintiff's expert witnesses in the 1984 New York census undercount trial.

Demographers have made an important contribution to small area research through modeling and statistical procedures. Verma and Basavarajappa describe the recent developments in applications of regression methods in estimating population for small areas with special reference to Canada. Among various approaches, the authors favor a multiple model framework for evaluating competing estimation techniques. In order to assess the relative accuracy of the alternative methods, a test was carried out, and its results are reported in the paper.

McCullagh and Zidek have developed a regression model for postcensal estimation of the population series of small areas. Their model was obtained by an axiomatic approach; its relationship to competing regression models is discussed in this paper. As far as evaluation criteria are concerned, the authors note that the performance of an estimation methodology requires answers to two questions: (1) Against what should the answers produced by the methodology be compared? (2) By what criterion is the comparison made?

Part 3, "Theoretical Developments," groups together four papers: they are by Dempster and Raghunathan; Fay; Fuller and Harter; and Stroud.

From the theoretical point of view, the development of efficient small area estimates involves modeling and associated statistical procedures. Among the earlier methods, synthetic estimation has been prominent and popular for some time. In fact, the first symposium on small area estimation held in 1979, sponsored by the National Institution of Drug Abuse, was entitled "Synthetic Estimates for Small Areas." Estimates based on synthetic methods utilize the assumption that small areas have the same characteristics as the larger areas of which they are a subset. More sophisticated estimators, relating to the original synthetic technique and exploiting complex association structure, have also been proposed. The so-called SPREE method is one example.

These early attempts marked an important realization among samplers that model-based reasoning may be necessary to come to grips with the special challenges posed by small area estimation.

Many further developments have followed involving considerable sophistication, for example, Bayes, empirical Bayes, shrinkage-type estimators, and regression estimators (model-based and randomization-based). A number of papers in this book reflect this important theoretical progress.

Dempster and Raghunathan's "common sense" Bayesian approach involves a case study based on partially hypothetical data. They address the problem of estimating total wages paid for each small area, using gross business income as a supplementary variable. Several Bayesian analyses were performed, using one-way random effect models with Gaussian errors.

The paper by Fay proposes an approach to the joint estimation of several small area parameters, involving multivariate linear regression. More specifically, a mixed components-of-variance model is assumed, leading to a "shrinkage-type" estimator for the vector of small area parameters. Broadly speaking, the procedure is "parametric empirical Bayes." Fay indicates how the procedure may be implemented in the calculation of small area estimates for the Current Population Survey carried out monthly by the Census Bureau.

The paper by Fuller and Harter proposes a multivariate regression model with components-of-variance error structure. They obtain nearly minimum mean square error predictors for small area means. Nearly unbiased estimates of the prediction mean square error are also constructed, under the assumption of Gaussian errors. An application to crop acreage estimation, using Landsat satellite data, is also mentioned.

The paper by Stroud develops a hierarchical Bayesian approach wherein the prior distribution of small area means contains hyperparameters that are subjected to a relatively diffuse prior distribution. Assuming that the prior means are related to a known auxiliary variable, Stroud derives exact formulas for the posterior means and posterior variance of small area means, for the case of equal sample sizes and Gaussian distributions.

Part 4, "Organizational Experiences with Small Area Estimation Techniques," consists of six papers from the following organizations: (1) The Australian Bureau of Statistics (by Feeney), (2) Statistics Canada (by Dagum, Hidiroglou, and Morry), (3) Statistics Sweden (two papers, the first by Cassel, Kristiansson, Råbäck, and Wahlström, and the second by Lundström), (4) the U.S. Bureau of the Census (by Isaki, Schultz, Smith, and Diffendal), and (5) the U.S. Bureau of Labor Statistics (by Cronkhite).

In his paper, Feeney points out that estimating the number of unemployed, based on a Monthly Labour Force Survey (monthly sample of households), is costly, in terms of dollars and respondent burden, in order to provide detailed information for specific small areas. To minimize the cost, he uses unemployment data from two sources, unemployment benefit recipients as well as the Monthly Labour Force Survey. The technique involves the use of an iterative fitting algorithm known as structure preserving estimation (SPREE) method. The performance of the model used by Feeney was found very satisfactory for all variables involved.

Monte Carlo simulation is an important element in the study by Dagum, Hidiroglou, and Morry. They examine the impact of classification and conceptual differences between a file based on a sample and a universe file of auxiliary variables. The performance of such methods as synthetic, ratio synthetic, regression count, and regression ratio estimators is discussed in their paper.

The two papers from Statistics Sweden deal with applications of small area techniques in Sweden. With a population of about 8 million, the country is, for administrative purposes, divided into roughly 280 municipalities. Reliable statistics for decision making concerning these low municipal levels are now urgently needed.

In estimating the unemployment for each municipality, a powerful source of auxiliary information in Sweden is provided by the information on job applicants registered with the national employment agency. This information is not based on a sample survey but can be used, with other auxiliary information, to improve the monthly survey estimates of unemployment. The paper by Cassel, Kristiansson, Råbäck, and Wahlström investigates various possibilities for building models and associated estimators in this context.

The paper by Lundström addresses the question of estimating, at the municipal level, the number of nonmarried cohabiting persons. For planning purposes such information is needed on a continuous basis, but is at present almost completely lacking for years between censuses. The paper uses a Monte Carlo simulation to compare various design-based and model-based estimators.

Isaki, Schultz, Smith, and Diffendal of the U.S. Bureau of the Census, dealing with Census undercount, focus their discussion on small area research in three directions. The first is to look at the result of the 1980 Post Enumeration Program. The second is the results of demographic analysis for 1980, and the third direction is referred to as adjustment methods. They also comment on the method advocated by Ericksen and Kadane, which involves the application of Bayesian hierarchical regression models for adjusting the census.

In his paper, Cronkhite discusses a method referred to as "The Handbook on Estimating Unemployment." Due to its major deficiencies, this method is replaced by single-equation regression models for estimating state employment and unemployment and pooled cross-sectional time series for estimating substate areas. In his critical assessment of various possible tests of model objective, Cronkhite points out the specific difficulties associated with each model objective.

Finally, Part 5, "Panel Discussion," is composed of the introductory comments offered by Dalenius, Kalton, and Kish.

In his remarks, Dalenius suggests a classification of estimation techniques and discusses the choice of standard for accuracy for small area estimates: Should the measure be a mean square error for an individual area estimate or some average of these as an overall measure of accuracy? Also, some potential topics for research and development are outlined.

The remarks by Kalton focus on the use of model-dependent estimators either by themselves or in combination with sample-based estimates. He emphasizes the need for evaluation studies to check on the adequacy of the models implicitly or explicitly being used. He recommends a cautious approach to the use of small area estimates, especially to their publication by government statistical agencies.

Kish in his discussion stresses the importance of classification and terminology for various subpopulations in future theoretical developments. He emphasizes the need for better auxiliary data and discusses briefly some approaches to small area estimation. He concludes that success depends first on better data and second on better methods. Finally he provides a number of references dealing with small area methodology.

The organization of the Symposium was due to the efforts of many persons. The local arrangement committee consisted of Jack Graham, Charles Patrick, Frank Mayda, and Gill Murray. We are particularly indebted to Frank and Gill for their excellent work in this regard. Also we wish to thank Dula Edirisinghe for her dedication and careful preparation of typing work.

We are grateful to Martin B. Wilk, the Chief Statistician of Canada at the time, who officially opened the Symposium, and to I. P. Fellegi, G. Kalton, L. Kish, G. Brackstone, and T. Dalenius for serving as session chairmen. The Panel Discussion provided a forum for expressing and exchanging many useful views and comments, and we thank T. Dalenius, G. Kalton, and L. Kish for acting as Panel Members at short notice. Finally, our thanks are due to Natural Sciences and Engineering Research Council of Canada for a Conference Grant and to Statistics Canada for providing further financial support.

R. PLATEK
J. N. K. RAO
C. E. SÄRNDAL
M. P. SINGH

Ottawa, Ontario, Canada
Montréal, Québec, Canada
October 1986

Contents

PART 1. POLICY ISSUES

Small Area Data: Policy Issues and Technical Challenges   1
G. J. Brackstone

PART 2. POPULATION ESTIMATION FOR SMALL AREAS

Sensitivity Analysis of Local Estimates of Undercount in the
1980 U.S. Census   23
E. P. Ericksen and J. B. Kadane

Recent Developments in the Regression Method for
Estimation of Population for Small Areas in Canada   46
R. B. P. Verma and K. G. Basavarajappa

Regression Methods and Performance Criteria for Small
Area Population Estimation   62
P. McCullagh and J. V. Zidek

PART 3. THEORETICAL DEVELOPMENTS

Using a Covariate for Small Area Estimation: A Common
Sense Bayesian Approach   77
A. P. Dempster and T. E. Raghunathan

Application of Multivariate Regression to Small Domain
Estimation   91
R. E. Fay

The Multivariate Components of Variance Model for Small
Area Estimation   103
W. A. Fuller and R. M. Harter

xiii

Bayes and Empirical Bayes Approaches to Small Area
Estimation   124
T. W. F. Stroud

PART 4. ORGANIZATIONAL EXPERIENCES WITH SMALL
AREA TECHNIQUES

Using Model-Based Estimation to Improve the Estimate of
Unemployment on a Regional Level in the Swedish Labor
Force Survey   141
*C.-M. Cassel, K.-E. Kristiansson, G. Råbäck, and
S. Wahlström*

Use of Regression Techniques for Developing State and
Area Employment and Unemployment Estimates   160
F. R. Cronkhite

Sensitivity of Small Area Estimators to Misclassification and
Conceptual Differences of Variables   175
E. B. Dagum, M. A. Hidiroglou, and M. A. Morry

The Estimation of the Number of Unemployed at the Small
Area Level   198
G. A. Feeney

Small Area Estimation Research for Census Undercount—
Progress Report   219
C. T. Isaki, L. K. Schultz, P. J. Smith, and G. J. Diffendal

An Evaluation of Small Area Estimation Methods: The
Case of Estimating the Number of Nonmarried Cohabiting
Persons in Swedish Municipalities   239
S. Lundström

PART 5. PANEL DISCUSSION

T. Dalenius   257
G. Kalton   264
L. Kish   267

Author Index   273

Subject Index   277

Small Area Statistics

PART 1

Policy Issues

Small Area Data: Policy Issues and Technical Challenges

G. J. Brackstone
Statistics Canada

ABSTRACT

This paper addresses issues arising in the development and provision of geographically integrated small area data to meet an increasing volume of user needs. From the perspective of a national statistical agency, issues are illustrated by tracing the conception and early development of Statistics Canada's Small Area Data Program. This Program embraces a wide variety of activities concerned with data development, data organization and dissemination, and the provision of the required infrastructure for these activities. The paper identifies policy issues related to this Program, and describes the various management and technical challenges imposed by such a Program.

1. INTRODUCTION

Interest in small area data seems to have grown markedly in recent years. Why should this be so? After all, small areas have existed for a long time, and small area data for almost as long. References to small area data, implied or explicit, can be found throughout history. Such undertakings as the Domesday book in 11th century England or the recording of baptisms, marriages, and burials in parish churches are rich with examples. Canada has a long history of small area data based on censuses dating back to 1666. The Census of 1667 conducted by Jean Talon in what was then known as New France revealed, for example, 291 families in the Québec area including 108 in Beaupré, 89 on Ile d'Orléans, 62 in Quebec

3

itself, and 32 at Beauport. Moving to 1851 and Hamilton, Ontario, we have available a range of statistical data including tables relating wealth and home ownership, and economic status and occupation, based on local assessment records (Katz, 1969).

The common feature of all these early examples is that they are all based either on a Census or on administrative procedures or records. In either case they depend on a process that aims at complete enumeration. Clearly the potential to produce small area data by such means has long been present; however, resources and technology have only allowed this potential to be realized intermittently.

The past 40 years have seen the emergence of sample surveys as a recognized means of meeting statistical data needs in many countries. Sample surveys have been very successful in supplying a wide range of national and subnational statistical data at frequent intervals and have become an important component, if not the pivotal component, of many countries' statistical programs. The drawback of sample surveys compared to censuses has, of course, been their inability to support reliable small area estimates. Thus, while statistical data bases at the national and subnational levels were becoming richer, the development of small area statistical data, except from periodic censuses, received scant attention.

The recent renaissance of interest in small area data can be traced to both demand and supply factors. Data users and analysts, their quantitative appetites whetted by richer national data bases, naturally seek further detail to extend their understandings. While more detail on a subject-matter topic may imply more, or more detailed, survey questions, extra detail in a geographic sense means small area data. The impetus for this increasing analytic demand comes from several sources. For example, there is in Canada, and probably in other countries too, an increasing government concern with issues of distribution, equity and disparity. An acceptable level of economic and social development at the national and subnational levels is seen as a necessary but not sufficient objective for national well-being. The existence of regions, or subgroups of the population, that are not keeping up with the average are causes for concern and remedial action. The need to identify such regions, to develop and simulate policy solutions, and to administer and evaluate the effects of programs on such regions, all imply a need for statistical data at the relevant geographic levels. In addition to these requirements generated by national governments, provincial and local governments continue to require data for their own areas for the purpose of analysis, planning, and administration within their jurisdictions. Furthermore, many business decisions, particularly decisions for small business, are made in the context of local social, economic, and environmental conditions, thus generating demands for small area data from the private sector. Lastly, but perhaps most importantly, come requirements for local data from elected representatives. The interest of Parliamentarians in statistical information about the areas they represent, both for assessing the impact of proposed policies or legislation and for comparing the well-being of their constituents with those in other areas, also adds to the demand for small area data, as do the needs of municipal politicians.

On the supply side there has been both statistical and technical progress conducive to the development of small area data. Administrative systems are now almost universally automated. Rapid technological developments in the storage and speed of processing of large data sets have made the regular processing of large administrative files for statistical purposes a more viable proposition than it was in earlier decades. Administrative files that contain a detailed geographic code (e.g., a postal code) can provide a valuable direct source of statistical data. On the methodological side, there has been a great deal of work in recent years on modeling methods that allow small area estimates, based jointly on survey and other census or administrative records, to be produced below the level that could be supported directly by the survey sample size.

The objective of this chapter is to provide an overview of the many issues confronting a statistical agency in trying to satisfy demands for small area data. We hope that this overview of issues might help to provide a framework or backdrop for the more specific and technical papers that follow. While it is clear from a number of papers presented at this Symposium that the concentration of statistical interest is on the problem of small area estimation, and while resolution of these problems is necessary for the production of small area data, we hope to illustrate in this paper that there are other interesting statistical problems arising in the domain of small area data, as well as a variety of nonstatistical issues.

To provide a focus for this paper we will begin by outlining the development and content of Statistics Canada's Small Area Data Program (SADP), and will then use this example to illustrate the range of policy, management, and technical issues that have to be faced.

2. STATISTICS CANADA'S SMALL AREA DATA PROGRAM

2.1. Program Conception

Prompted primarily by increasing demands for small area data among federal departments, Statistics Canada, in 1982, began developing a proposal for systematic and integrated development and dissemination of small area data. Before that time, small area data were being produced and disseminated by some individual subject-matter divisions (e.g., Agriculture or Construction). These efforts were naturally viewed from the perspective of the particular subject-matter topic and were regarded as a further breakdown of statistical data on that topic. Except perhaps for the Census of Population, there was no program that viewed small area data from a geographic perspective whereby the small area was the unit to be described and the different subject-matter topics represented the "detail" of that description.

In formulating its proposed program, Statistics Canada stressed several comparative advantages that it felt it possessed in taking on such an initiative. These included:

1. Its existing programs of small area data collection (particularly Census).
2. Its access to administrative records under the Statistics Act, and its program to develop administrative data for statistical purposes.
3. Its role as custodian of the country's Standard Geographic System (SGC) and supporting tools—particularly postal code to small area conversion files.
4. Its mature network of data dissemination systems including particularly CANSIM, which was seen as one vehicle for small area data dissemination.
5. Its expertise in statistical methods, which was seen as essential for developing techniques of small area estimation.

Given these programs, facilities, and expertise, the SADP was conceived as a program that would integrate and build upon Statistics Canada's existing small area data activities to provide a continuing, coherent output of small area data relevant to user needs. It was argued that a moderate additional investment would have high leverage by building upon existing infrastructure. It would provide a focus for small area data issues within Statistics Canada and would serve to sensitize program managers to the need to consider small area data needs in program design. Furthermore, the SADP was seen as a means of coordinating other foci of small area statistical work that existed in other federal departments (e.g., Department of Regional and Industrial Expansion, Environment Canada).

Interest in small area data was not restricted to the federal government. Clearly provincial, regional, and local governments have an interest in small area data within narrower geographic domains. Several provincial statistical agencies were already active in small area data work.

A Federal–Provincial Committee on Small Area Data, with membership from the statistical agencies of all the Provinces and Territories, was formed in 1982 and contributed to the development of the program proposal.

Late in 1982 a draft proposal was circulated to federal departments and provincial and territorial governments for comments. Reaction was very supportive and a subsequent submission for funding from the Federal Government received approval in 1983—but at a lower level of funding than requested. This funding was for a 3-year developmental period after which the Program was to become largely self-sustaining. The SADP was formally launched in July 1983.

2.2. Program Content

The proposed SADP had three major components—data development, data systems, and infrastructure—each aimed at one of the three objectives of the Program:

1. To produce new small area data sets to meet important user demands.
2. To organize and make available small area data in geographically oriented data bases and through supporting products.
3. To put in place geographic, conceptual, and methodological frameworks and tools needed to support continuing small area data development, dissemination, and analysis.

(a) Data Development

The acquisition of data to fill recognized and important small area data needs was seen as a primary component of the SADP. However, further extension of Statistics Canada's survey and Census programs was not regarded as a realistic approach to meeting small area data needs. While Censuses will remain as a source of periodic benchmarks for small area data, and sample surveys will continue to provide more frequent national and provincial benchmarks, considerations of cost and respondent burden preclude further significant expansions in these areas. The solution must be sought elsewhere. Four generic approaches to data development that are being pursued both separately and in combination can be identified:

1. Further tapping of existing census and survey programs for their full small area potential. (Improved geographic coding and development of annual averages from monthly surveys are two specific approaches in this category.)

2. Further development of administrative records for statistical purposes.

3. Acquisition of data from other jurisdictions.

4. Statistical modeling techniques using data from censuses, surveys, and administrative records as input.

(b) Data Systems

The objective of any statistical program is to put statistical information into the hands of users. In the case of small area data this constitutes a formidable problem given the volume of data involved, the dispersed sources of data, and the wide variety of users. A substantial part of the problem is knowing what small area data exist. Therefore, an important element in the SADP's data systems is the development of an inventory of existing small area data sets, including both their characteristics and information on how they can be accessed.

The remaining part of the problem concerns the method of making data available to users. The overall strategy being adopted by the SADP is illustrated in Figure 1. It is based on a conceptual division of small area data holdings into source data bases and summary data bases. Source data bases, shown on the right of Figure 1 may reside anywhere—usually with the collector of the information. Some will be in Statistics Canada,

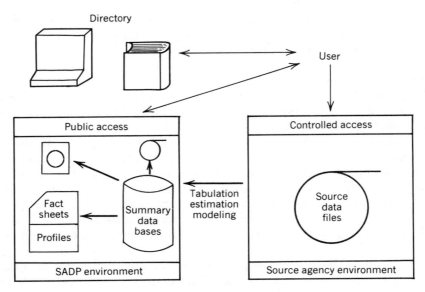

Figure 1. Small area data systems.

some elsewhere. They will often contain individual records or microdata, usually confidential, and therefore data retrieval from them will be subject to confidentiality protection procedures. Access to data will require contact with the custodian of the source data base, will often require a special retrieval request, and will therefore not be immediate. Source data bases will be referenced in the small area data directory. Only data bases from which the custodian is prepared to service retrieval requests should be included as source data bases.

Summary data bases, on the other hand, represent consolidations of data from a variety of sources at a single geographic level in a single data base. Their features are that they would be publicly and easily accessible; their contents will already have been screened for confidentiality; they should be supported by display and analytic facilities required by users; and they will relate to a prespecified system of geography. Feeding off summary data bases will be a series of standard products that could include factsheets (brief summary statistics on an area), profiles (more detailed statistical pictures of areas), data profiles on diskettes, or summary data tapes.

The data systems component of the SADP is concerned with building the facilities and assembling the data needed to implement the system. An incremental approach is being adopted with initial emphasis being placed on the creation of an inventory, certain summary data bases, and the development of statistical profiles both on paper and on diskette.

(c) Infrastructure

The activities of data development and data dissemination described in the previous components require frameworks and supporting services to operate. The most obvious and critical element of this infrastructure is the geographic framework. In a real sense "geography" is the "subject-matter" of the SADP. While the Standard Geographic Classification indeed provides a standard, it is not sufficient. The SADP must aim to provide data for areas other than just the standard areas. This implies an ability to tabulate microdata records for ad hoc areas. This in turn requires systems of geocoding that allow records to be associated with geographic areas small enough that flexible aggregation into user areas is possible. Two such systems exist. The first is geocoding, which involves the coding of latitude and longitude coordinates, and which has been used in Canadian Censuses of Population since 1971 (Statistics Canada, 1972). The second is postal coding, which utilizes the postal code assigned to a mailing address. Both of these systems generally distinguish block faces in urban areas, but become less precise in rural areas. An important element of the SADP is the further development of both these systems, particularly the development of postal code to geographic code conversion files. Such files are critical to the assignment of geographic codes to administrative record files which often contain postal codes.

A second element of the infrastructure component is the development of a subject-matter framework for small area data. This addresses the following question: What is the set of data items needed to describe a given type of small area? Such a question is unanswerable in that form, and yet some concept of the array of data items that constitute the "minimal sufficient statistic" for a particular small area data analysis problem does seem pertinent. A first step toward addressing this type of issue may be the development of a taxonomy of small areas, categories of which could include, for example, market towns, mining communities, transportation centers, dormitory towns, etc. (Berry, 1972).

Finally, under the title of infrastructure we include research and development activities into techniques needed to support the development of small area data. This element overlaps with Statistics Canada's existing program of methods research but should put particular emphasis on problems of small area estimation and confidentiality protection.

2.3. Current Status

What has just been described is the vision and general direction of the SADP; it has not yet been achieved. So far, data development has concentrated on labor market statistics and basis statistics on economic activity. These topics are addressed in other papers in this symposium. In the data systems component, some successful products have been produced including profiles for Electoral Districts and county data on diskettes for Ontario. A summary small area data base is currently being created in CANSIM, Statistics Canada's electronic dissemination system. The main activity in the infrastructure component has been the development of postal code conversion files. A preliminary version of a Canada-wide file that relates postal codes to the lowest realistic unit of the SGC (down to block face in urban areas) is being released this month.

3. POLICY ISSUES

Under this heading we consider two distinct issues: (1) government policy directions that influence either the demand for, or supply of, small area data and (2) issues of statistical policy that arise in the development and provision of small area data.

As mentioned earlier, increased emphasis on regional economic development and the impact of policies on subprovincial regions and small areas has stimulated demand for subprovincial data within the federal government. This demand has appeared both as an increase in ad hoc demands for small area data and in a few specific demands for regular systematic provisions of small area data for the administration of programs or the application of regulations. To illustrate these regular program uses, two particular applications will be described: the Regional Industrial Development Program (RIDP) administered by the Department of Regional and Industrial Expansion (DRIE) and the Unemployment Insurance Program administered by Employment and Immigration Canada.

The RIDP requires the calculation of a "development index" for each of 260 Census Divisions in Canada. This index is then used to rank Census Division into four tiers that qualify for successively higher maximum levels of government assistance for approved industrial development projects. Calculation of the index requires data on income levels, on unemployment, and on employment to be supplied annually for each Census Division. This application has raised a range of issues that have broader connotations. There is a fundamental question of how an index like this should be defined. Can statistical theory and methods help in defining which variables should go into the index and how they should be measured and combined? Exploration of that question would take us far beyond the scope of this paper. More pragmatically, there is a question of how the statistician should react to a requirement to provide data that are straining the limits of acceptable reliability for a formula that will lead to the allocation of funds. In this particular case, errors in data will generally only influence fund allocation for Census Divisions that fall close to the boundary of different tiers, and, even then, the index only leads to the determination of maximum assistance levels—it does not ensure these levels will be granted. Nevertheless, the statistician is faced with the problem of what to advise in situations where existing data cannot support the intended usage. A related question is the choice of geographic area for which to apply such an index. In this case the choice was heavily influenced by the availability of data with the result that the chosen regions may not be the best regions to utilize from the viewpoint of regional economic analysis. Some research into the definition of more appropriate regions has been undertaken.

The second example relates to the application of the Unemployment Insurance (UI) system in Canada. Regulations stipulate different qualification and requalification periods in different regions according to the unemployment rate in the region. The period of prior continuous employment required to qualify for unemployment benefits is shorter in regions in which the unemployment rate is higher. The unemployment rate specified for use in this determination is a 3-month moving average of seasonally adjusted estimates for so-called UI regions. These areas are large enough that the required estimates can be derived from the monthly Labour Force Survey. However, this use of local area estimates for determining benefit eligibility puts these figures in the spotlight. Local allegations that unemployment is being underestimated, thus depriving the local unemployed of benefits to which they are entitled, are not unusual. Such complaints tend to focus more on the conceptual basis for measuring unemployment—the active job search requirement—than on sampling error resulting from small sample sizes.

Turning to a separate aspect of public policy, the issues of privacy and access to information are pertinent to all statistical work, but particularly to a small area statistics program because of its reliance on administrative records. There appear to be two conflicting viewpoints prevalent in this debate. The first, based on a privacy argument, is that information provided to government should be used only for the purpose for which it was collected, with the respondent being notified of this purpose at the time of collection. This is the stipulation of Canada's Privacy Act (sections 5 and 7) with respect to personal information. However, statistical purposes are identified in the Act as one of the explicit permissible secondary uses. The other viewpoint, based on a respondent burden concern, is typified by the complaint "I already told the X Department that, why are you asking me again?" This viewpoint must suppose a far higher degree of data exchange between departments than in fact exists. Under the Statistics Act, Statistics Canada may have access to administrative records of other departments for statistical purposes, but it cannot allow data records it has acquired under the Statistics Act to be passed to anyone outside Statistics Canada.

The exploitation of administrative records for statistical purposes quickly leads one to consider the potential of record linkage between administrative systems. This is particularly the case if one has in mind the objective of replacing some part of a census or survey program with administrative data. No single administrative system rivals the completeness and data richness of a census or broad sample survey, so that combination of administrative records is essential if one is to construct a data base containing a useful range of characteristics. Given the public concern about privacy, databanks, etc., large-scale record linkage should not be initiated without careful consideration of the political as well as the technical implications. In its own development of administrative records to produce small area data, Statistics Canada has essentially dealt with separate administrative systems independently and has not engaged in extensive record linkage. Proposals for record linkage at Statistics Canada are now subject to a thorough review and approval process.

The discussion of privacy and record linkage conveniently moves us into the area of statistical policy. Two further statistical policy issues are reviewed in the following paragraphs.

A national small area data program has to be a collaborative effort. While the Canadian Constitution assigns responsibility to the Federal Government for statistical matters, this does not diminish the role of other levels of government both as users and suppliers of statistical data particularly for small areas. Provincial government responsibilities for major social program areas (e.g., health, education) result in them being custodians of important administrative records with considerable small area potential. Municipal governments are in a similar position with respect to other data sets. Furthermore, there exist commonalities of data requirements between federal and provincial governments which have the potential to lead to duplication in the absence of a collaborative effort. Clearly then, some mutual understanding of roles is necessary to achieve maximum benefit from limited dollars. In Canada, the Federal–Provincial Committee on Small Area Data is charged with responsibility for reaching such understandings.

We have alluded already to the issue of the quality of small area data. A significant problem in small area data development is the assessment of data quality. Coupling this with the fact that the reliability of small area data is often inferior to that of most other data outputs of a statistical agency, careful attention must be paid to policy on release of data of marginal quality and on apprising users of the quality of data they are receiving.

4. MANAGEMENT ISSUES

Under this heading we shall mention issues of program management, revenue generation, and priority determination in the context of a small area data program.

The management of the SADP presents some novel challenges in a statistical agency traditionally organized along subject-matter lines. By its nature, the SADP transcends subject-matter boundaries and its success is dependent on its ability to harness the cooperation of these subject-matter areas both in identifying existing small area data sets and in developing new data sets to meet user needs. Furthermore, the SADP depends on support from a variety of functional areas including Geography, Methodology, Informatics, and Marketing.

The program management approach adopted for the SADP in Statistics Canada involves a small line staff, headed by a Program Manager, and responsible for program planning and monitoring, product development, production, and marketing. This staff in turn contracts with over Divisions of the Bureau to undertake activities, particularly data development projects, in support of the SADP. An internal Steering Committee of senior managers provides guidance to the Program Manager and sets priorities between the various activities proposed for inclusion within the Program.

An important objective of the SADP is to become financially self-supporting after the initial 3-year developmental period. In order to continue the evolution of the Program beyond the developmental period, including, for example, investment in new data development activities, the Program aims to achieve revenues through the sale of products and services sufficient to support most of its continuing activities. This aim is consistent with Statistics Canada's general policies with respect to recovery of the marginal cost of packaging and making data available to users. However, it requires internal mechanisms to ensure that the appropriate portion of revenues is funneled back to the Program.

The issue of priority determination, mentioned earlier as a responsibility of the Steering Committee, is a difficult one. How should one choose between the competing demands for new data sets? Should data development be exclusively demand driven, or should consideration be given to the development of a balanced array of descriptor variables for small areas at some level of subject-matter and geographic detail? To date, the SADP's data development focus has been on meeting the major demands, particularly of federal departments, where these needs could be met using existing data sources and available statistical techniques. Thus emphasis has been on labor market and income data for individuals and families, and data on business activity (number of establishments, employment, gross revenue, etc.) by industry and size of business. These major areas have been supplemented by further development of population estimates. The idea of developing a framework for the integration of small area data remains on the research agenda.

5. TECHNICAL ISSUES

A program as diverse as the SADP raises a variety of technical issues, particularly in the areas of geography, statistical methodology, and data management and dissemination. In this section some of these issues are enumerated and analyzed without going too much into their technical details.

5.1. Geographic Structure

The SADP attempts to provide statistical data regularly for standard areas and on demand for nonstandard areas. This immediately raises two issues: how to choose the standard areas and how to provide the flexibility needed to produce data for nonstandard areas.

One could regard the problem of standard areas as resolved by the existence of the Standard Geographic Classification (SGC) in Canada. This provides a hierarchy of geographic areas, respecting certain administrative boundaries, and forms the basis for much of Statistics Canada's publication program, including the Census of Population. However, while the SGC has provided a fairly stable structure for dissemination and analysis of population-related data, some disadvantages have emerged with respect to its use for subprovincial economic data.

Specifically, for the purposes of regional economic analysis (including, for example, the administration of the RIDP mentioned earlier), areas that are functionally related by economic activities seem more relevant than areas delimited by administrative boundaries. In the case of economic data, much of which is business based, the flexibility to produce small area data for a variety of geographic systems is particularly limited due to problems of residual disclosure. Therefore, it is important to define in advance the most useful set of areas for which such data should be made available. Census Metropolitan Areas, which are based on a labor market concept, and which are an element in the SGC, may represent a partial solution to this problem. However, these do not cover the whole of Canada. The solution to this problem may lie in an extrapolation of this concept to cover the remaining parts of the country.

Flexibility in producing data for user-defined areas is best achieved by maintaining a data base of microdata in which each record carries a very detailed geographic code. This is the procedure adopted in the Census data base (where the geographic code is the pair of latitude–longitude coordinates), and in many administrative files (where the geographic code is the postal code). This approach requires a conversion file that relates the required areas (standard or nonstandard) to a set of the detailed geocodes or postal codes. Simple aggregation of records can then yield the required statistical data. The SADP is making widespread use of postal code conversion files. Issues that arise in this approach include the specificity of the detailed codes (e.g., postal codes in rural areas often refer to large tracts of land that receive postal service from a central office), and the fact that on some administrative files, the postal code may relate to a location other than the one for which the unit should be tabulated (e.g., an accountant's address as opposed to a home address on a tax record).

Characterizing the previous approach as "aggregation," we can identify an alternative disaggregation approach. In situations where data are available only for large geographic areas, or only for a set of areas different from those required, we may nevertheless be able to convert the data to the required areas by some form of distribution or disaggregation usually based on distribution information for a related variable (e.g., a previous census or a highly correlated current variable). This takes us into the realm of statistical estimation problems that fall under the heading of statistical methodology.

5.2. Statistical Methodology

As we indicated earlier, direct estimation from censuses or sample surveys is not a practical option for meeting the full range of small area data demands. Therefore, the development of methods of estimation that combine existing data from censuses, sample surveys, and administrative records using statistical models is a critical component of the SADP. The range of methods investigated includes synthetic estimation; regression models, including sample regression techniques (Ericksen, 1974); structure preserving ratio estimation (Purcell and Kish, 1980); sample-dependent estimation (Drew, Singh, and Choudhry, 1982); and other variations and combinations of these approaches (Statistics Canada, 1983). Since many of these approaches are discussed in detail in other chapters, they will not be pursued further here.

A general problem arising in utilizing most of these methods is the problem of evaluation. How can one assess the quality of the resulting estimates? This is a real problem for a statistical agency that has to decide whether or not to issue particular estimates, and, if so, with what caveats. This problem of quality evaluation is not unique to small area estimates but is magnified and pervasive in this area. Approaches that have been used include detailed evaluations of the quality of administrative records, cross-validation studies, and simulation of estimation procedures using census data. Periodic census data also play a fundamental role in the evaluation and recalibration of models. There is perhaps a more fundamental question of what we mean by quality, and how we should characterize quality, for estimates that utilize modeling methods. For survey-based estimates, sampling error measures are normally supplied. These are extended in some cases to cover bias and variance arising from nonsampling sources. In the case of model-based estimates, it is usually a model variance that is quoted. Often, in the small area context, this variance represents an average measure of reliability over many small areas. In other cases, for example, in the System of National Accounts or in demographic estimates, it may be a "statistical discrepancy" or an "error of closure" that is quoted. We run the risk of confusing users who may be woefully indifferent to quality measures anyway.

On the issue of small area data evaluation, it is also worth noting that error in small area estimates may be more apparent to users than error in national aggregates. Few users are in a position to question national or even provincial estimates, but at a local level there will be critics quick to point out deficiencies. Of course, such feedback can be positive and lead to improved estimation procedures, but, nevertheless, it is true that for small areas, where estimation is more difficult, scrutiny of estimates is also more intensive.

Another issue present in most statistical dissemination activities, but heightened in the case of small area data, is the problem of confidentiality protection—ensuring that data about individual respondents cannot be inferred from published tabulations. This problem is more acute for business data than for individual data because of the smaller numbers of units and the problem of dominance of aggregates by a few large units. The various confidentiality protection techniques that have been developed, including random rounding, perturbation, and suppression (Nargundkar and Saveland, 1972; Cox, 1980), are all useful. However, the problem of combining flexibility in tabulation with protection of confidentiality, particularly for quantitative as opposed to frequency data, requires further research.

Finally, under statistical methodology, we should refer to the technical problems of record linkage. As mentioned earlier the linkage of different administrative files, or of administrative files with survey files, can be an important element in small area data development. While in some cases, we may be dealing with unique identifiers as a basis for linkage, there will always be at least a residue of cases for which a statistical decision framework is required for determining when a match is deemed to exist. A statistical record linkage system, CANLINK, has been developed at Statistics Canada although it has been used primarily for mortality or morbidity studies rather than small area applications.

5.3. Data Management and Dissemination Issues

Two distinctive characteristics of small area data and their use influence heavily the methods by which small area statistics are stored and disseminated. The first characteristic is the voluminous nature of small area statistics, compared to national and provincial statistics. Even a limited array of statistical data for each of many Census Divisions, electoral districts, or municipalities will cover many printed pages. The second characteristic relates to the typical users of small area data. While there are users, particularly at the federal government level, who are seeking comparative data for a complete set of small areas spanning the whole country, the interests of most users are restricted to a relatively small subset of small areas.

These two factors lead to a bipartite dissemination strategy that involves (1) storage of the full range of available small area statistics in such a way that selected data for selected areas can be extracted quickly and efficiently on demand and (2) prepackaging of certain key statistics for a whole set of small areas into profiles that may be published in print or made available on diskette. Production of data on demand would be on a cost recovery basis; prepackaged profiles would carry a predetermined price.

A range of issues arise in attempting to create and market packaged products. For example:

1. On what basis to make the selection of variables for inclusion.
2. To what extent, and how, to provide the descriptive information on data sources, methodology, and quality that should accompany statistical data.
3. How to handle problems of inconsistency in definitions or concepts, classification systems, or reference periods, in different data sets brought together in a profile.

These are all issues that exist within a single statistical program but which are exacerbated in a program that cuts across existing statistical programs. For example, there are business survey programs that utilize different vintages of the Standard Industrial Classification. This may be tolerable for the analyst working with the output of a single survey, but difficulties arise when one tries to integrate data across surveys. In some instances, the mere act of juxtaposing data for the same area from different sources may serve to identify data problems in one or other source, thus yielding benefits beyond the confines of small area data.

6. CONCLUSION

We have come this far without defining what we mean by a small area. This is no oversight. Our concept of a small area, for those who prefer logical definitions, can be understood by applying a transitive rule of logic to the two well-known clichés: "Small is beautiful" and "beauty lies in the eye of the beholder." A small area is what a user perceives to be a small area, and a federal bureaucrat in Ottawa will have a different perception than a local planner in Calgary. The qualifier "subprovincial" has been used to help circumscribe the scope of the SADP, but even this term conjures up quite different areas in Prince Edward Island than in Ontario. Of course, some standard small areas have to be defined for creating summary data bases and other products as mentioned earlier, and Census Divisions, municipalities, and Electoral Districts have been used as standards so far. However, we are not adopting a definition that restricts small areas to a particular set of standard areas.

An alternative, but equally imprecise, approach to defining a small area is to regard as a small area any area for which direct design-based estimates cannot be reliably produced from the current sample survey program, or a reasonable expansion thereof. Small areas thus become areas that need other methods of estimation. This definition will be attractive to many contributors to this symposium as it corresponds closely to the statistician's concept of the small area estimation problem. From the lay-user's viewpoint, this definition is void of meaning; he or she is more interested in numbers than methods of estimation.

It is appropriate to end there, where we began, with users' needs. Users want numbers they can use, and they want to feel that someone, or some agency, is standing behind those numbers. Their interest in methods of estimation is at best secondary (although it can become primary in confrontations over the accuracy of numbers).

In this chapter we have tried to stress the practical aspects of the provision of small area data, while also touching on some of the theoretical and methodological problems that exist in this field. While problems of small area estimation are clearly central and important components of a small area data system, we hope that this chapter will have helped to put these problems into a broader perspective and to stimulate interest in other issues arising in a program of small area data development and dissemination.

ACKNOWLEDGMENTS

The author wishes to acknowledge his colleagues at Statistics Canada whose thought and work has served to develop many of the ideas and issues described in this chapter. However, responsibility for the way in which these ideas and issues have been summarized and presented rests with the author.

REFERENCES

Berry, B. J. L. (Ed.) (1972), *City Classification Handbook: Methods and Applications*, Wiley-Interscience, New York.

Cox, L. H. (1980), "Suppression Methodology and Statistical Disclosure Control," *Journal of the American Statistical Association*, **75**, 377–385.

Drew, J. D., Singh, M. P., and Choudhry, G. H. (1982), "Evaluation of Small Area Estimation Techniques for the Canadian Labour Force Survey," *Survey Methodology*, **8**, 17–47.

Ericksen, E. P. (1974), "A Regression Method for Estimating Population Changes of Local Areas," *Journal of the American Statistical Association*, **69**, 867–875.

Katz, M. B. (1969), "Social Structure in Hamilton, Ontario," in *Nineteenth Century Cities, Essays in the New Urban History*, Yale University Press, New Haven, CT.

Nargundkar, M. S., and Saveland, W. (1972), "Random Rounding: A Means of Preventing Disclosure of Information about Individual Respondents in Aggregate Data," *Proceedings of the Social Statistics Section of ASA*.

Purcell, N. J., and Kish, L. (1980), "Postcensal Estimates for Local Areas (or Domains)," *International Statistical Review*, **48**, 3–18.

Statistics Canada (1972), *GRDSR: Facts by Small Areas; the Geographically Referenced Data Storage and Retrieval System, an Introduction*.

Statistics Canada (1983), "A Bibliography for Small Area Estimation," *Survey Methodology*, **9**, 241–261.

PART 2

Population Estimation for Small Areas

Sensitivity Analysis of Local Estimates of Undercount in the 1980 U.S. Census

E. P. Ericksen
Temple University

J. B. Kadane
Carnegie-Mellon University

ABSTRACT

We have used a hierarchical Bayesian model to compute local estimates of the undercount in the 1980 U.S. Census. This chapter analyses the sensitivity of these estimates to variations in the assumptions on which they are based. These assumptions concern the numbers and racial composition of undocumented aliens, strategies for imputing values to missing data in the survey on which the estimates are based, and methods of computing standard errors. This chapter also investigates the problem of extrapolating to areas other than those on which the model is estimated.

1. INTRODUCTION

Our objective is to compute local area estimates of population undercount in the 1980 U.S. Census. The method we study here for doing so is to obtain sample estimates of the undercount for a set of local areas, and then to use regression to relate these to a set of predictor variables. For areas with sample data, we compute final weighted averages of the sample and regression estimates; for other areas, we simply apply the regression equation.

We originally presented our method as plaintiffs' expert witnesses in the New York census undercount trial (*Cuomo v. Baldrige*, pp. 2141–2468, but see pp. 2496–2880 for Census Bureau defense). While the method has much in common with previous research (Ericksen, 1974; Carter and Rolph, 1974; Fay and Herriot, 1979; Morris, 1983; Du-Mouchel and Harris, 1983), the context in which we made our presentation has led to an unusual amount of commentary and debate [e.g., Ericksen and Kadane (1985) and Freedman and Navidi (1986), each with comments and rejoinder]. This provides an opportunity to discuss technical issues pertinent to the local estimation problem in general. We will focus attention on four specific problems:

1. Missing values required imputation for 8–10% of cases in the sample survey on which the estimates are based. Different imputation strategies led to different results, and an arbitrary choice of one strategy may cause the final estimates to be arbitrary as well.
2. Millions of undocumented aliens live in the United States, and they are not included in the demographic estimate of the population total. Some argue that this renders impossible the estimation either of national or local undercount rates.
3. The estimated variances of the local sample estimates may be unreliable, causing the final weighted averages to be unreliable as well.
4. Areas outside the set of local areas with sample estimates differ from the sample areas, and this may make it difficult to use the regression equation to make extrapolations.

In this paper, we propose to evaluate these four arguments.

Our analysis was based on data collected in the Post Enumeration Program (PEP) of the 1980 Census, which had two components. One, known as the P-sample, included the samples for the April and August Current Population Surveys (CPS), and these were matched against census records to estimate *omissions*, people who should have been counted but were not. The other component, known as the E-sample, was an independently selected sample of census households that were revisited to verify the correctness of the count and therefore to measure *erroneous enumerations*, people who were counted more than once or who should not have been counted. These components were then combined to provide estimates of the undercount for states, large SMSAs, and cities. Our analysis includes 66 of these—16 cities, 12 remainders of states in which they are located, and the remaining 38 states.

In the discussion that follows, we first present our estimation model, which is based on the work of Lindley and Smith (1972). Next, we describe the problem of missing data on the PEP and how demographic information can be used to minimize its import. Following this we address the four specific criticisms. This is done by varying the assumptions on which our estimates are based and then looking to see whether the estimates have changed in important ways. We then finish with a short conclusion.

2. THE REGRESSION MODEL

In an earlier article, we (Ericksen and Kadane, 1985) proposed a local estimation model incorporating an average of a local sample estimate and a systematic estimate based on a regression equation. In this application, the data are expressed in terms of the measured undercount, $y_i = 100(1 - X_{c,i}/x_{s,i})$, where $X_{c,i}$ is the count for area i and $x_{s,i}$ is the sample population estimate for area i. The y_i can be either positive (undercount) or negative (overcount). Suppose γ_i is the true undercount in area i, and that τ_i^2 is the sampling variance for y_i. Then, using sampling theory alone, we have

$$y_i \sim N(\gamma_i, \tau_i^2), \qquad i = 1, \ldots, I \tag{1}$$

independently. Equation (1), taken alone, suggests that one might estimate γ_i, the true undercount in area i, by y_i, the sample estimate for each of the 66 areas. It is well known that such sample estimates are likely to exaggerate unduly differences among areas (James and Stein, 1961). Consequently, it is appropriate to pull these estimates toward some common value. To pull them in entirely would be to adjust every area by the same amount, which would have no effect on apportionment or fund allocation. Thus, we need a model that will permit the data to indicate an appropriate amount of smoothing. A simple and tractable model that permits smoothing is

$$\gamma_i \sim N(\beta, \sigma^2), \qquad i = 1, \ldots, I \tag{2}$$

independently. As $\sigma^2 \to 0$, each γ_i is, regardless of the data, brought toward the common value β. Conversely, as $\sigma^2 \to \infty$, the posterior mean for each γ_i more and more nearly approaches the initial data y_i. Thus the system (1) and (2) can be used to represent a wide variety of beliefs about appropriate adjustment. This system is incomplete, however, in the following respect. Some places may reasonably be expected to have greater undercounts than other places, for systematic reasons, like the proportion of its population minority, whether or not it is a large central city, and whether or not the conventional method of census taking was used. Equation (2) does not take these auxiliary facts about area i into account, but rather treats each place identically. This line of thought leads to replacing equation (2) with

$$\gamma \sim N(X\beta, \sigma^2 I) \tag{3}$$

where X is a matrix of dimension $I \times p$, β is a p-vector, and γ is the I-dimensional vector $(\gamma_1, \ldots, \gamma_I)$. Equation (3) is a generalization of equation (2) as long as the matrix X includes a column of 1's. To recapitulate the argument from before, the view that no adjustment is appropriate is equivalent to $\beta = (1, 0, \ldots, 0)$ and $\sigma^2 \to 0$. The view that the PEP data should be taken literally and used without smoothing is equivalent to $\sigma^2 \to \infty$. Values of σ^2 in between are compromise positions. Finally, the view that the undercounts are determined totally by a model (3) is equivalent to $\sigma^2 \to 0$.

3.   MISSING VALUES IN THE SURVEY DATA

In equation (1), we assumed that the sample estimates, y_i, were normally distributed about the true undercounts, γ_i. If, however, there should be a third term θ_i, such that the y_i were distributed about $\gamma_i + \theta_i$, and if the θ_i took on substantial values, then we would be estimating something other than the undercount, such as error in the PEP data. This can be brought about when the count/omit status of an unmatched CPS person is unresolved, that is, we cannot tell whether the CPS person was counted or not. Relying on different assumptions on how unresolved cases should be treated, the Census Bureau produced 12 series of PEP estimates, which did vary somewhat. This suggests that each series j produced a set of θ_{ij}, at least some of which are substantially different from zero. To estimate local undercounts, we must either select a series where we are confident that the θ_{ij} are small, or we must show empirically that using different PEP series leads to the same results, meaning that the θ_{ij} are similar.

Our next step is to describe the PEP method and to explain the different missing data strategies used by the Census Bureau. The procedure for determining omissions included three steps. First, the CPS sample households were matched against the Census rolls. If anyone in the CPS household appeared to be missed, that person's record was sent to "followup," which meant that the Census Bureau tried to find and interview this person in late 1980. The purpose of the interview was to determine the person's whereabouts on Census Day to make sure that the person really had been omitted. In about 75% of the followup cases, a successful determination of the count/omit status of the CPS person was made. The remaining 25% were *unresolved* by followup and a status was imputed by various strategies (Cowan and Bettin, 1982).

The problem of making correct determinations also bedeviled the E-sample, but to a lesser extent. Here, 2.8% of the sample was unresolved after initial interviews. For about one-third of these, pertinent information was obtained from the Post Office, but for the remainder, an enumeration status had to be imputed. Brief descriptions of the P- and E-sample imputation strategies now follow.

3.1. Omissions

There are five series of omissions estimates, three based on the April CPS (and labeled Series 2, 3, and 14) and two based on the August CPS (and labeled Series 5 and 10).

Series 2: This is the *linking* alternative. Unresolved cases were linked to resolved followup cases having similar characteristics (e.g., race and sex). The count/omit status of the linked resolved case was assigned to its matched unresolved case. This is similar to the imputation strategy used on the Decennial Census.

Series 3: This is very similar to Series 2, except for the treatment of "type A noninterviews." These are cases were a CPS interview was obtained in March or May, and attempted but not completed in April. In Series 2, the March or May information was used to provide a basis for matching to the census, while in Series 3 such cases were omitted from the analysis.

Series 14: This is the *found person* alternative, in which unresolved cases were simply omitted from the analysis. This is tantamount to assuming that omission rates for resolved cases are the same as omission rates for unresolved cases, whether the resolved cases were sent to followup or not. Since most resolved cases were matched to the census, this strategy produced substantially lower omission rates than did the strategies for Series 2 or 3.

Series 5: This is the *August* analog to Series 2, but incorporating one additional feature. For persons who had moved between April and

August, it was necessary to obtain the April address. For cases where the April address could not be obtained, imputations were made that did not take the geographical location into account.

Series 10: In this case, movers were simply omitted from the analysis, whether the April address was found or not. This had the practical effect of assuming that movers had the same omission rates as nonmovers and thus lowered the estimate.

3.2. Erroneous Enumerations

There were two strategic decisions made for unresolved E-sample cases. First, the Census Bureau asked the Post Office to provide information as to whether the person had actually lived at the address indicated for April, 1980. After some of these designations had been obtained from the Post Office, the Census Bureau became suspicious about their accuracy. The Bureau therefore calculated one set of estimates, *Series 8*, in which the Post Office information was used, and another set, *Series 9*, in which it was discarded. Second, unresolved cases for which Post Office information was not obtained were linked to resolved cases, and imputations were made in the manner of P-sample Series 2, 3, and 5. The third E-sample set, *Series 20*, omitted the unresolved cases from the analysis, and they were assumed to have the same erroneous enumeration rate as the resolved cases.

The Census Bureau combined the P- and E-sample data to produce 12 of the 15 logically possible series of estimates. These are designated by two numbers with, for example, Series 2-8 consisting of omission Series 2 and erroneous enumeration Series 8.

3.3. Choosing Among the 12 Series

There are logical reasons to prefer estimates based on omissions Series 2, 3, and 5 to those based on Series 10 and 14. It seems illogical to omit movers from the August sample, and unreasonable to assume that unresolved April cases are like the full sample. After all, a census record was found for the large majority of resolved cases, but for none of the unresolved cases. We shall see in Section 3.4 that estimates based on series 10 and 14 are less consistent with demographic analysis—they seriously underestimate the Black undercount.

We also note that some of the series are very similar. Comparing estimates with the same E-sample assumptions (e.g., 2-8 with 3-8), we find that correlations between Series 2 and 3 are all between .99 and 1.0. Similarly, correlations between E-sample Series 9 and 20 are all about

.99. Consequently, to simplify our presentation, we have eliminated Series 3-8, 3-9, 2-20, 3-20, and 14-20, and we will focus attention on the remaining seven series. Among these, national undercount rates vary from -1.0 to $+2.0\%$, and there are differences among the local distributions. To deal with this problem, we have introduced demographic information.

3.4. Calibrating PEP to a Demographic Total

The dual-systems model for estimating the undercount, used by the Census Bureau in its analysis of the PEP data (Cowan and Bettin, 1982), is shown in Table 1. In the PEP application, n_1 is the number of persons counted in the census, minus erroneous enumerations and persons imputed into the count, $n_{.1}$ is the estimated number of persons covered by the CPS, and n_{11}, n_{12}, and n_{21} are all estimated from the matching study. The quantity Z is

$$Z = kn_{12}n_{21}/n_{11} \qquad (4)$$

In calculating its PEP estimates, the Census Bureau arbitrarily set k equal to 1.0, and the quantity T was obtained by adding the estimated Z to the observed n_{11}, n_{12}, and n_{21}. Because the CPS misses important parts of the population (U.S. Bureau of the Census, 1978, p. 62), and because the assumptions underlying the different PEP series vary, the estimates of Z and T also vary, and a criterion for choosing among these estimates and correcting for the undercoverage of the CPS is needed.

Our proposal is to use demographic analysis to provide values of T for Blacks, non-Black Hispanics, and all others, and then to solve for Z and k. These values of T will be used for all PEP series, having the effect of equalizing *national* undercount rate estimates among the series. This strategy would not, of course, affect the rank ordering of the local undercount rate estimates. Our strategy can be expressed as follows. If

Table 1. Population Distribution in Dual Systems Estimation Model

	Alternative List		
Census	Yes	No	Total
Yes	n_{11}	n_{12}	$n_{1.}$
No	n_{21}	Z	$n_{21} + Z$
Total	$n_{.1}$	$n_{12} + Z$	T

we write P_{ij} as the PEP estimate for group j in area i, T_{cj} as the Census Bureau's PEP estimate of the national total for group j, and T_{dj} as the corresponding national demographic estimate, our strategy is to compute, for each local area i,

$$P_i = \sum_j (T_{dj}/T_{cj})P_{ij} \qquad (5)$$

(a) Calculating National Totals

We start with the demographic estimates computed by the Census Bureau (1982), and add estimates of Black and non-Black undocumented aliens. We then subdivide the non-Black total by assuming that the non-Black Hispanic undercount rate is equal to that of Blacks, and dividing the counted number of non-Black Hispanics by one minus this rate. The estimated number of "All Others" is obtained as a residual, and the corresponding undercount rate is obtained by comparison with the census count.

In 1980, Census Bureau demographers Siegel, Passel, and Robinson (1980, p. 19) reviewed existing studies of the numbers of undocumented aliens and concluded that:

> The total number of illegal residents in the United States for some recent year, such as 1978, is almost certainly below 6.0 million, and may be substantially less, possibly only 3.5 to 5.0 million.

In a later study, Bureau demographers Warren and Passel estimated that 2.047 million was a lower bound estimate of the number of undocumented aliens counted in the Census (Warren and Passel, 1983). This conservative estimate assumed, among other things, that no legal aliens had been omitted from the Census, and that citizenship status had been reported accurately. These two studies lead us to conclude that the number of undocumented aliens in the United States is between 2 and 6 million, and that 4 million is a reasonable estimate.

Warren and Passel also estimated the numbers of undocumented aliens coming from each of 40 regions, and they stated that approximately 109,000 came from the largely Black Caribbean countries of Jamaica, Haiti, and Trinidad and Tobago, and that 93,000 came from Africa and Oceania. At least some of the undocumented aliens coming from other countries such as the Dominican Republic are likely to be Black, so we conclude that 8–10% of the 2 million counted undocumented aliens are Black. We will assume 8%. [Passel testified that this is a reasonable conclusion to be derived from their analysis in *Cuomo v. Baldrige* (p. 1018).]

Depending on the PEP series, the survey data indicate that the Black undercount rates are slightly higher or slightly less than the Hispanic rates. The differences are small enough perhaps to be accounted for by random sampling variation, suggesting that the undercount rates may be substantially equal. In our analysis, we will make the assumption that they are.

When we add the 4 million undocumented aliens to the demographic estimate of total population computed by the Census Bureau, the overall undercount rate becomes 1.4%, and the rates for Blacks and Hispanics are 5.8%. The residual rate for the non-Hispanic, non-Black remainder is 0.3%.

(b) Calibrating the PEP Results

The calculated ratio, T_{dj}/T_{cj}, are shown in Table 2. Ratios less than 1.0, as for Blacks and Hispanics in Series 2-9, indicate that k is less than 1.0. Ratios greater than 1.0 indicate that k is also greater than 1.0. The closeness of the ratios to 1.0 for Series 2-8, 2-9, 5-8, and 5-9 indicate that there is good agreement between the demographic estimates and PEP, and this helps to justify our preference for these series over 10-8, 14-8, and 14-9. As for the other four series, there is little difference among them, although we find the assumptions underlying Series 2-9 to be more reasonable than those underlying the other three.

Table 2. Calibration Ratios Aligning PEP with Demographic Estimates[a]

| | Group | | |
Estimate	Blacks	Non-Black Hispanics	All Others
2-8	1.002	1.002	1.002
2-9	0.990	0.989	1.000
5-8	1.019	0.982	0.994
5-9	1.007	0.970	0.991
10-8	1.037	1.012	1.007
14-8	1.056	1.052	1.016
14-9	1.043	1.039	1.014

[a] National totals for Blacks and non-Black Hispanics are based on "modified" counts as given in U.S. Bureau of the Census (1982).

4. REGRESSION ANALYSIS FOR 66 AREAS

We now turn to estimating the regression equation. Our objective was to find a parsimonious set of independent variables that led to small values of $\hat{\sigma}^2$ and regression coefficients that were at least twice their standard errors. Eight variables, available for each of the 66 areas, were considered. They are:

1. Percentage of population Black or Hispanic (% minority).
2. Numbers of crimes per 1000 population (crime rate).
3. Percentage of census population counted by the conventional (no mailout–mailback) method.
4. Percentage of population living in poverty.
5. Percentage of population having difficulty speaking English (language difficulty).
6. Percentage of adults not high school graduates.
7. Whether or not the area was a large central city (16 yes, 50 no).
8. Proportion of housing in small, multiunit structures.

All subsets of two, three, or four of these variables were regressed against all seven of the PEP series. Consistent results were obtained, with the three-variable subset including % minority, crime rate, and % conventional (items 1–3) giving the best result in almost every case. Differences were often close, and equations substituting the language difficulty or central city variables for crime rate were almost as good as the three-variable equations we picked. However, omitting any of the three variables led to increased values of $\hat{\sigma}^2$.

Matrix weighted averages of the regression and sample estimates were computed and these are shown in Table 3. The averages are

$$d = (D^* + (1/\sigma^2)\bar{P}_X)^{-1}D^*y \qquad (6)$$

where D^* is an $n \times n$ diagonal matrix with ith diagonal element $1/\tau_i^2$ and where $\bar{P}_X = 1 - X(X'X)^{-1}X'$ (Ericksen and Kadane, 1985, p. 105). The sample sizes for most areas were between 1000 and 2000, but the reliance on a clustered design within these areas led to values of τ_i^2 that were somewhat higher than would be expected under assumptions of simple random sampling.

For ease of presentation, we have grouped the 66 areas into four categories: (A) large central cities with many minorities, (B) areas with 10–39.9% minority, (C) areas with fewer than 10% minority and less use of the conventional method, and (D) areas with less than 10% minority and greater use of the conventional method. The conventional method, where no prior address list was compiled and enumerators personally collected the census information, was used in the most rural parts of the

Table 3. Final Estimates, $\hat{\gamma}_i$, of Population Undercount for 66 Areas

Area	PEP Series Used							Ranges	
	2-8	2-9	5-8	5-9	10-8	14-8	14-9	First Four[a]	All Seven
Group A: Central Cities with at Least 40% Black or Hispanic									
Baltimore	4.9	5.2	4.6	4.9	4.1	4.1	4.4	0.6	1.1
Chicago	3.2	3.5	3.3	3.4	3.0	2.8	3.1	0.3	0.7
Cleveland	4.4	4.8	4.2	4.6	3.7	3.5	3.9	0.6	1.3
Dallas	5.2	5.8	4.4	5.0	3.9	3.2	3.7	1.4	2.6
Detroit	5.3	5.8	5.1	5.7	4.6	4.3	4.7	0.7	1.5
Houston	3.4	3.7	3.4	3.6	3.0	2.4	2.6	0.3	1.3
Los Angeles	4.5	5.2	3.7	4.3	3.5	3.4	4.3	1.5	1.8
New York	4.8	5.2	3.7	4.2	3.5	3.5	3.9	1.5	1.7
Philadelphia	2.7	2.7	2.4	2.4	2.3	2.9	2.9	0.3	0.6
Saint Louis	5.9	6.7	5.5	6.4	4.8	4.0	4.7	1.2	2.7
Washington, DC	5.4	5.9	5.2	5.8	4.8	4.7	5.3	0.7	1.2
Group B: Areas with 10–39.9% Black or Hispanic									
Alabama	0.6	0.4	0.8	0.7	1.2	1.0	0.8	0.4	0.8
Arizona	2.4	2.7	2.7	2.9	2.4	2.1	2.3	0.5	0.8
Arkansas	−0.2	−0.4	0.4	0.2	0.5	0.7	0.4	0.8	1.1
California(R)[b]	2.8	2.9	2.1	2.2	2.0	2.5	2.6	0.8	0.9
Colorado	1.6	1.8	2.1	2.3	2.0	1.2	1.4	0.7	1.1
Connecticut	0.4	0.4	0.6	0.6	0.7	0.7	0.6	0.2	0.3
Delaware	0.8	0.8	1.3	1.4	1.4	1.1	1.1	0.6	0.6
Florida	2.1	2.2	2.3	2.4	2.0	1.7	1.8	0.3	0.7
Georgia	0.8	0.7	1.5	1.4	1.4	1.0	0.9	0.8	0.8
Louisiana	1.6	1.5	1.7	1.6	1.7	1.8	1.8	0.2	0.3
Maryland(R)	1.2	1.1	1.3	1.2	1.2	1.6	1.5	0.2	0.5
Mississippi	0.9	0.7	1.4	1.1	1.3	1.8	1.6	0.7	1.1
Nevada	2.6	3.3	2.3	2.7	2.0	1.8	2.5	1.0	1.5
New Jersey	1.5	1.3	1.3	1.2	1.3	1.7	1.5	0.3	0.5
New Mexico	3.1	3.1	3.7	3.8	3.8	2.7	2.6	0.7	1.2
North Carolina	0.9	0.8	1.1	0.9	1.1	1.6	1.4	0.3	0.8
South Carolina	1.8	1.7	1.7	1.6	1.6	2.1	2.1	0.2	0.5
Tennessee	−0.4	−0.5	0.3	0.1	0.6	0.3	0.2	0.8	1.1
Texas(R)	0.9	0.9	1.6	1.6	1.6	0.8	0.8	0.7	0.8
Virginia	0.7	0.6	1.0	0.9	1.1	1.3	1.2	0.4	0.7
Boston	4.7	5.6	4.3	5.0	3.8	2.8	3.6	1.3	2.8
Indianapolis	1.0	0.8	1.3	1.2	1.3	1.1	0.9	0.5	0.5
Milwaukee	2.1	2.1	1.8	1.8	1.8	2.3	2.2	0.3	0.5
San Diego	2.0	2.3	2.1	2.3	2.0	1.8	2.0	0.3	0.5
San Francisco	3.6	4.1	3.1	3.5	2.7	2.5	2.9	1.0	1.6

Table 3. (*Continued*)

Area	2-8	2-9	5-8	5-9	10-8	14-8	14-9	First Four[a]	All Seven
			PEP Series Used					Ranges	

Area	2-8	2-9	5-8	5-9	10-8	14-8	14-9	First Four[a]	All Seven
Group C: Areas with Less than 10% Black or Hispanic, Less than 75% Conventional									
Hawaii	1.9	2.0	1.9	2.1	2.1	1.5	1.5	0.2	0.6
Idaho	1.4	1.3	2.0	1.9	2.0	1.7	1.6	0.7	0.7
Illinois(R)	1.1	1.0	0.0	−0.1	0.4	1.1	1.0	1.2	1.2
Indiana(R)	−0.2	−0.4	0.4	0.2	0.5	0.3	0.1	0.8	0.9
Iowa	−0.3	−0.5	0.0	−0.2	0.1	0.4	0.2	0.5	0.9
Kansas	0.8	0.7	1.0	0.9	1.0	1.0	0.9	0.3	0.3
Kentucky	−0.8	−1.1	−0.2	−0.6	0.0	−0.1	−0.4	0.9	1.1
Maine	1.2	1.1	0.9	0.8	1.2	1.4	1.2	0.4	0.6
Massachusetts(R)	−0.5	−0.3	−0.3	−0.2	0.3	0.1	0.2	0.3	0.8
Michigan(R)	0.8	0.8	0.5	0.5	0.8	1.1	1.1	0.3	0.6
Minnesota	0.8	0.9	0.4	0.3	0.7	1.4	1.4	0.6	1.1
Missouri(R)	0.4	0.3	0.0	−0.2	0.3	1.2	1.0	0.6	1.4
Nebraska	0.4	0.3	0.8	0.7	1.1	1.1	1.0	0.5	0.8
New Hampshire	−0.6	−0.8	−0.1	−0.3	0.0	0.0	−0.2	0.7	0.8
New York(R)	−1.1	−1.3	0.2	0.0	0.4	−0.3	−0.5	1.5	1.7
Ohio(R)	0.8	0.6	0.4	0.2	0.6	1.1	0.9	0.6	0.9
Oklahoma	0.2	0.0	0.3	0.2	0.5	0.0	−0.2	0.3	0.7
Oregon	0.8	1.0	1.1	1.2	1.1	0.6	0.8	0.4	0.6
Pennsylvania(R)	−0.8	−0.9	−0.7	−0.9	−0.3	0.1	0.0	0.2	1.0
Rhode Island	0.7	0.5	0.5	0.4	0.5	0.9	0.6	0.3	0.5
Utah	0.7	0.9	1.0	1.0	1.0	1.3	1.5	0.3	0.8
Vermont	−0.5	−0.8	−0.3	−0.5	−0.1	0.3	0.0	0.5	1.1
Washington	1.3	1.2	1.2	1.2	1.1	1.1	0.9	0.1	0.4
West Virginia	−1.0	−1.3	−0.2	−0.7	−0.5	0.3	−0.1	1.1	1.6
Wisconsin(R)	1.0	0.7	−0.3	−0.5	0.4	1.7	1.5	1.5	2.2
Group D: Areas with Less than 10% Black or Hispanic, 75–100% Conventional									
Alaska	3.0	3.2	3.5	3.7	3.7	2.7	2.9	0.7	1.0
Montana	1.7	1.7	2.0	2.1	2.4	1.6	1.6	0.4	0.8
North Dakota	0.4	0.3	0.8	0.8	1.7	1.1	1.0	0.5	1.4
South Dakota	0.7	0.6	1.6	1.5	2.1	1.1	1.0	1.0	1.5
Wyoming	2.7	2.8	3.0	3.1	3.3	2.6	2.8	0.4	0.7

[a] "First four" refers to estimates produced by Series 2-8, 2-9, 5-8, and 5-9.
[b] (R) refers to "remainder of state," after listed central cities have been removed.

United States, especially in the West. Its use appears to have led to lower than average rates of erroneous enumeration but somewhat higher rates of omission. Looking now at Table 3, we see a clear pattern. In Group A, the highest undercount rates are found, and they are usually between 3 and 6%. In Group C the lowest rates are found, usually between −1 and 2%. Intermediate rates are found in Groups B and D, usually between 0 and 3%.

We also see that the estimates are consistent when different PEP series are used for the same area. For the "best four" series (2-8, 2-9, 5-8, 5-9) the range is less than 1.0% in 54 of 66 cases and less than 2.0% in all cases. Looking at all seven estimates, the range is less than 1.0% in 37 cases, between 1.0 and 1.9% in 25 cases, and between 2.0 and 2.9% in 4 cases. In general, selection of a PEP series makes little difference, and we can conclude that, for the most part, values of θ_{ij} are inconsequential. We can also conclude that a series of adjustments that followed the general pattern of the PEP series, with the largest increases in Group A, the smallest increases in Group C, and intermediate increases in population in Groups B and D, would improve the estimated distribution of population in the United States.

5. SENSITIVITY OF ESTIMATES TO CHANGES IN DEMOGRAPHIC ASSUMPTIONS

We made important assumptions concerning numbers and characteristics of undocumented aliens and the undercount rates of Blacks and Hispanics. The sensitivity of the estimates to reasonable modifications in the assumptions needs to be evaluated. To do this, we recalculated estimates for Series 2-8 with the following changes in assumptions: (1) the numbers of undocumented aliens were set equal first to 2.0 and then to 6.0 million, (2) the proportions of undocumented aliens who were Black were set equal first to 6% and then to 10%, and (3) the Black–Hispanic differential in undercount rates was set equal to ±2%. We then adjusted the sample estimates to sum to the new national totals T_{dj}, and summed the totals across the Groups A, B, C, and D as shown in Table 4. The resulting group level undercount rates are shown for Series 2-8, but essentially the same conclusion would be drawn for any of the other PEP series.

Of the 27 sets of undercount estimates, the overall rates are lower when the number of undocumented aliens is set equal to 2 million and higher when the number is 6 million. Holding the number of undocumen-

Undocumented Aliens		Black–Hispanic Undercount Rate Differential (%)	Group				Differences Between Areas		
Number (millions)	Percentage Black		A	B	C	D	A–D	D–B	B–C
2	6	+2	4.5	0.5	−0.5	0.9	3.6	0.4	1.0
		0	4.7	0.6	−0.6	0.8	3.9	0.2	1.2
		−2	5.0	0.6	−0.7	0.6	4.4	0.0	1.3
	8	+2	4.6	0.5	−0.5	0.9	3.7	0.4	1.0
		0	4.8	0.6	−0.6	0.7	4.1	0.1	1.2
		−2	5.0	0.6	−0.8	0.6	4.4	0.0	1.4
	10	+2	4.6	0.5	−0.5	0.8	3.8	0.3	1.0
		0	4.9	0.6	−0.7	0.7	4.2	0.1	1.3
		−2	5.1	0.7	−0.8	0.6	4.5	−0.1	1.5
4	6	+2	5.2	1.3	0.5	1.8	3.4	0.5	0.8
		0	5.4	1.4	0.3	1.7	3.7	0.3	1.1
		−2	5.6	1.5	0.2	1.6	4.0	0.1	1.3
	8	+2	5.3	1.4	0.4	1.8	3.5	0.4	1.0
		0	5.5	1.4	0.3	1.7	3.8	0.3	1.1
		−2	5.8	1.5	0.2	1.5	4.3	0.0	1.3
	10	+2	5.4	1.4	0.4	1.7	3.7	0.3	1.0
		0	5.6	1.4	0.3	1.6	4.0	0.2	1.1
		−2	5.9	1.5	0.1	1.5	4.4	0.0	1.4
6	6	+2	5.9	2.2	1.4	2.8	3.1	0.6	0.8
		0	6.1	2.2	1.3	2.7	3.4	0.5	0.9
		−2	6.3	2.3	1.2	2.5	3.8	0.2	1.1
	8	+2	6.0	2.2	1.3	2.7	3.3	0.5	0.9
		0	6.2	2.3	1.2	2.6	3.6	0.3	1.1
		−2	6.5	2.3	1.1	2.5	4.0	0.2	1.2
	10	+2	6.2	2.2	1.3	2.6	3.6	0.4	0.9
		0	6.4	2.3	1.2	2.5	3.9	0.2	1.1
		−2	6.6	2.4	1.0	2.4	4.2	0.0	1.4

[a] Undercount rates were computed by summing the PEP values, T_{cij} across the areas in the four groups, adjusting these by the ratio, T_{dj}/T_{cj} implied by the demographic assumptions made, and comparing these to the population counts. The Black undercount rate was held constant, and the Black–Hispanic differentials computed by changing the assumed Hispanic rates.

ted aliens constant and varying their proportion Black or the overall Black–Hispanic differential seem to cause only small variations.

Turning to the between-group differences, we see a consistent pattern. Group A always has the highest undercount, followed by Groups B and D, with Group C always having the lowest undercount. The same pattern was observed in the regression estimates, and this reinforces our major conclusion, that interarea differences are stable. In Table 4, the A–D difference is always between 3.1 and 4.5%, the D–B difference is always between 0.6 and −0.1%, and the B–C difference is always between 0.8 and 1.5%. The maximum difference, A–C, always falls between 4.5 and 5.9%. Again, we see that an undercount adjustment shifting population from Group C to A, with smaller adjustments made for Groups B and D, would improve the estimated population distribution. Varying the demographic assumptions has a moderate effect on our estimates of T, Z, and k (Table 1), but only small effects on the between-area differences in the undercount rates. Because these variations span the range of reasonable assumptions, we conclude that for any set of reasonable assumptions concerning numbers of undocumented aliens, their racial composition, and Black–Hispanic undercount differentials, we would come to about the same conclusions concerning which areas had large and which areas had small undercounts.

6. POSSIBLE EFFECTS OF ERRORS IN ESTIMATED SAMPLING VARIANCES

Freedman and Navidi (1986), in a paper based on Freedman's testimony in the New York lawsuit, suggest that errors in the estimation of sampling variances, τ_i^2, might spoil the weighted average, or posterior estimates, $\hat{\gamma}_i$. It is surely the case that we do not have exact values of the τ_i^2, as these are measured with error like any other sample estimate. Sampling variations alone might cause some areas to have variances estimated too low and other areas to have variances estimated too high. Moreover, if the estimated variances exclude important sources of nonsampling error, there might be a systematic factor causing these estimates to be too low on average.

If their argument is correct, errors in the estimation of τ_i^2 could have three important effects: (1) the 66 individual area estimates would not be properly weighted when calculating the regression equation, since the areas are weighted in inverse proportion to the τ_i; (2) the regression and sample estimates would not be properly weighted in calculating the posterior estimate, $\hat{\gamma}_i$; and (3) estimates of σ^2 would be incorrect.

The problem can perhaps be made more concrete with an example. Consider the cases of New York City and South Carolina under PEP Series 2-8. The New York City sample estimate was .066 (an undercount of 6.6%) with a standard error (τ_i) or .014. Corresponding estimates for South Carolina were .060 and .029. When regression estimates based on % minority, crime rate, and % conventional were calculated, New York's estimate was .042 and South Carolina's .014. The regression and sample estimates were more consistent for New York City than for South Carolina.

The fact that the New York sample estimate had a smaller standard error meant that it received greater weight in the calculation of its posterior estimate, $\hat{\gamma}_i$, than did South Carolina's sample estimate when its posterior was calculated. The New York City posterior was .048, 25% of the distance from the regression to the sample estimate. The South Carolina posterior was .018, 9% of the distance from the regression to the sample estimate. Finally, the fact that New York City, where the sample estimate and regression estimate were more consistent, received greater weight helped to reduce the value of $\hat{\sigma}^2$, giving the appearance of a satisfactory goodness of fit.

This is a common problem in local estimation, recognized by Ericksen (1974), among others. Ericksen observed that most of his local sample estimates were derived from similarly sized samples, so the expected sampling errors would be approximately equal. Consequently, he assigned equal weights to these areas, and for the remaining areas with larger sample sizes, larger weights were assigned. This reduced the importance of extreme cases with very small sampling variances and increased the importance of extreme cases with very large variances.

To evaluate Freedman and Navidi's argument, we made two experiments. The first, consistent with Ericksen's strategy, was to substitute "median" estimates for the individual τ_i. Because the sample sizes among the 50 states and state remainders did not vary greatly, the median τ_i among these 50 was computed and assigned to each area. A similar procedure was applied to the 16 central cities, where sample sizes were smaller. Consequently, one set of weights was assigned to the states and another, smaller, set of weights was assigned to the cities.

The second experiment attempted to evaluate the possibility that the Census Bureau estimates of τ_i were too low. We arbitrarily doubled each of the Census Bureau estimates, which had been calculated by half-sample replication methods, and thus increased the τ_i^2 by a factor of four. In each of these experiments, we looked to see if the posterior estimates. $\hat{\gamma}_i$, had changed in important ways, or if the standard errors of the posteriors had been increased. Such an increase would indicate that we were exaggerating our confidence in the posterior estimates based on Census Bureau variance estimates.

In Table 5, we present distributions of changes obtained in $\hat{\gamma}_i$ when three different assumptions are used. The assumptions are

1. Sampling variances, τ_i^2, computed by Census Bureau using half-sample replication methods are correct.

2. For the 50 states and state remainders, a common, "median," value is used, with a separate common median used for the 16 central cities. They are

PEP Series	2-8	2-9	5-8	5-9	10-8	14-8	14-9
States	.009	.009	.010	.010	.009	.008	.008
Cities	.020	.020	.019	.019	.017	.016	.015

3. Original τ_i were multiplied by 2.0.

There we can see that changing the values of τ_i had little effect on the $\hat{\gamma}_i$. In the large majority of cases, the difference was less than 0.5% and in 95% of the cases, the difference was less than 1%. When comparing estimates based on the "median" rather than the original estimates of τ_i, the largest changes occurred for areas like South Carolina where the original value of τ_i was large. Replacing assumption 1 with assumption 2 caused the posterior estimate for South Carolina to increase from 1.8 to 4.6%. This is much greater than the undercount estimate obtained in any other southern state. In general, it seems to be a good idea to heed the

Table 5. Distributions of Differences, $\hat{\gamma}_i - \hat{\gamma}_i'$, is Posterior Estimates When Assumptions Concerning Sampling Variances Are Changed

Assumptions Being Compared	Difference, $\hat{\gamma}_i - \hat{\gamma}_i'$	PEP Series						
		2-8	2-9	5-8	5-9	10-8	14-8	14-9
1–2	0.0–0.49	61	63	51	52	63	62	63
	0.5–0.99	3	2	13	12	3	1	1
	1.0–1.99	2	1	2	2	0	2	1
	2.0–2.99	0	0	0	0	0	1	1
1–3	0.0–0.49	52	54	61	63	66	48	47
	0.5–0.99	11	8	5	3	0	17	18
	1.0–1.99	3	4	0	0	0	1	1
	2.0–2.99	0	0	0	0	0	0	0

signal given by a large value of τ_i and not to assign that sample estimate very much weight.

We now turn to the effects of the changed τ_i on the standard errors of the $\hat{\gamma}_i$. The medians of these standard errors for estimates based on assumptions 1, 2, and 3 are shown in Table 6. Starting with the comparison of estimates based on the first two assumptions, we see that there were scarcely any changes in the standard errors of $\hat{\gamma}_i$. We see that the standard errors, compared to undercount estimates ranging as high as 6%, are small and consistent.

Turning to the comparison of estimates based on the first and third assumptions, we observe the perhaps surprising result that the standard errors of the $\hat{\gamma}_i$ were actually *reduced*. This result can be understood, though, by recalling that the observed y_i consists of three components: (1) a 'systematic and explained" part reflecting the variation in the three variables included in the regression equation, (2) a "systematic and unexplained" part reflecting variation in variables not included in the regression equation, and (3) random variation attributable to sampling error. When we arbitrarily increased the estimated τ_i, this had the effect of reducing the second component, and caused our three explanatory variables to appear to become much more powerful predictors. Indeed, for Series 2-8, $\hat{\sigma}^2$ was an unrealistically low .002 (reduced from .53). The posterior estimates $\hat{\gamma}_i$ were pulled more strongly toward the regression estimates when the τ_i were increased. Since the original estimates were already rather close to the regression estimates, this caused little change in the $\hat{\gamma}_i$. In the latter case, the apparently more accurate regression estimates were given greater weight, and this caused the posterior estimates to appear to be more accurate as well. In conclusion, then, we see that making rather substantial changes in the nature of the sampling variances τ_i^2 have had little effect either on the distributions of the posterior estimates, $\hat{\gamma}_i$, nor on their standard errors. While it will be helpful if future Census Bureau research can increase our confidence in

Table 6. Median Standard Errors Obtained for 66 Area Estimates of γ_i, According to Three Assumptions Concerning τ_i^2

Assumption	PEP Series						
	2-8	2-9	5-8	5-9	10-8	14-8	14-9
1	.64	.64	.50	.48	.33	.58	.59
2	.64	.64	.63	.64	.45	.60	.60
3	.43	.42	.46	.46	.42	.36	.36

the accuracy of the τ_i, the possibility that currently available estimates might be wrong should not influence our views concerning the credibility of our posterior estimates nor the feasibility of adjustment in any important way.

7. EXTRAPOLATING TO LOCAL AREAS WHERE SAMPLE DATA ARE NOT AVAILABLE

Freedman and Navidi raised the additional issue of whether it would be possible to extrapolate to areas other than the 66 where sample data are available. They proposed the following test of robustness. Because crime rates are known to be higher in large central cities, and because these central cities are 100% urban by the Census Bureau definition of "urban," they suggested that it was reasonable to expect that substituting "percentage urban" for the crime rate variable in the regression equation also including the percentage of minority and conventional would lead to similar results. They showed that this expectation was generally borne out for the 66 areas on which the regression equation was based. They then identified 12 counties, most of which had very small populations, having either high crime rates (over 100 per 1000 population) and no urban population or low crime rates (below 5) and highly urban populations. Plugging values of independent variables into the regression equations, they showed that the regression estimates for these 12 counties varied widely (1–4%). They concluded that, because such seemingly similar sets of independent variables gave such disparate results, the stability of extrapolations was questionable, and at least one of the regression estimates likely to be wrong.

We were suspicious of their results for two reasons. One is that Freedman and Navidi's 12 counties were small and unrepresentative. Second the urbanization variable takes on a very different meaning when applied to large central cities than to adjacent suburbs or small towns that are also 100% urban by the Census Bureau definition. That is why we did not include the urbanization variable in the set of eight predictors of undercount tested for inclusion in our regression equations. Even so, it is still possible that extrapolations are questionable, and we replicated their method for a more representative set of areas.

We selected a probability sample of 28 central cities and remainders of SMSA, and 46 nonmetropolitan counties. Our sampling method was systematic, within alphabetical lists of SMSAs and counties, the latter sorted by state. We calculated two regression equations, and took the difference, as follows:

$$\hat{Y}_1 = -2.46 + .048\text{Min} + .045\text{Crime} \qquad\qquad + .024\text{Conv}$$
$$\hat{Y}_2 = -3.22 + .058\text{Min} \qquad\qquad + .045\text{Urban} + .027\text{Conv}$$
$$\hat{D} = \quad 0.76 - .010\text{Min} + .045\text{Crime} - .045\text{Urban} - .003\text{Conv}$$

$$(7)$$

where Crime is the crime rate, and Min, Urban, and Conv are the percentages minority, urban, and enumerated by the conventional method. Differences, $\hat{D}$, were calculated for each of the 66 PEP areas and for each of the $(28 + 28 + 46) = 102$ areas in the systematic sample. All calculations were based on PEP Series 2-8 and results are shown in Table 7.

We can see that most of the discrepancies are small, and the problem cited by Freedman and Navidi is not very common. Nearly three-quarters, 72%, of the discrepancies are less than 1.0, and only 11, 6%, are greater than 2.0. $\hat{D}$ exceeds 4.0 in two cases, Aleutian Islands, Alaska and Daytona Beach, Florida. Both areas have high crime rates, 183 per 1000 in Daytona Beach and 156 in the Aleutian Islands where only 43% of the population is urban. Equation (7) shows that $\hat{D}$ will be large when the crime rate is very different from the percentage urban, so these results are not surprising.

Comparing groups of areas, we see that the smallest discrepancies are obtained in the 50 states and state remainders. This is to be expected because (1) they received over 90% of the weight when the regression equations were calculated and (2) they are heterogeneous areas without extreme values in the independent variables. The crime rates in these 50 areas range from 25 to 88 per 1000 and the percentage urban ranges from 34 to 89%.

Larger errors obtain in central cities, whether they are PEP areas or

Table 7. Distributions of Differences in Regression Estimates for Local Areas

	Type of Area				
	66 Areas on which Regression Equation Was Calculated		Systematic Sample of 102 Other Areas		
Absolute Percentage Difference	16 Central Cities	50 States and Remainders	28 Central Cities	28 Remainders of SMSA	46 Nonmetropolitan Counties
0–0.9	9	49	14	20	29
1.0–1.9	5	1	10	7	13
2.0–2.9	2	0	3	1	1
3.0 or more	0	0	1	0	3
Median	0.7	0.3	1.0	0.7	0.9

selected into the systematic sample. The central cities, urban by the Census Bureau definition, had a wider range in crime rates. Large values of $\hat{D}$ were obtained when crime rates were very high, as in Saint Louis (143) or Daytona Beach, or when crime rates were very low, as in Kokomo, Indiana (41) or Glen Falls, New York (39).

Turning to the suburban and nonmetropolitan areas, the results are similar. The two cases were $\hat{D}$ was between 3.0 and 3.9 were Clare and Oscoda Counties, Michigan. In the first case, the crime rate was 85 and the percentage urban was 14% and in the second case the crime rate was 64 and the percentage urban was 0.

The fact that large discrepancies occurred rarely does not mean that we should be unconcerned about it, though. All areas need adjustments, and we can tailor our strategy to reduce the problem. We have three observations and a design recommendation to make.

First, a careful choice must be made of the independent variables in the regression equation. "One hundred percent urban" has a very different meaning in New York City (included in calculating the regression equation) than in Bowling Green, Kentucky (not included), and this can only be known on the basis of *a priori* knowledge and subjective judgment. When the meanings of key variables differ inside and outside the sample areas, difficulties in extrapolation can be expected.

Second, we must pay attention to the quality of data used for the independent variables. If we have reason to believe that crime data are questionable for some areas, it would be better to use a different equation with a better-measured set of independent variables. It is likely that some of the discrepancies observed both by us and by Freedman and Navidi are due to poor quality crime data.

Third, many of the areas where crime rates are especially high are those where the residential population is quite different from its "de facto" population much of the year. Daytona Beach, Florida is a vacation area where many vacationers come for temporary visits. They are properly excluded from the census count in Daytona Beach. These vacationers, though, can be perpetrators or victims of crimes. The crime rate, then, applies to a larger population than that the Census Bureau attempts to count, and it may be inappropriate for vacation areas. This seems to be especially true for Alpine County, California, cited by Freedman and Navidi. This is a resort area in the Sierra Nevada mountains where the crime rate was 252 per 1000. Similar problems could occur in areas with "special" populations, like large military bases, with many young adult males.

These points pertain to unusual areas. More importantly, extrapolations were made difficult in this application because the regression equation was calculated on a set of generally heterogeneous areas with moderate crime rates and percentages minority. Many of the areas for which estimates are required have crime rates and percentages minority much higher or lower than found in the 50 states receiving most of the weight in calculating the regression equation. This problem must be attended to by revising the design of the PEP. Rather than using states as units of observation, it would be much better to create more homogeneous areas, some with high and some with low crime rates and percentages minority. Care should also be taken to include areas with low crime rates and many minorities and high crime rates and few minorities. If this is done, then the regression equation will be calculated on a set of areas which more nearly reflect the variation existing among those areas where estimates are needed. This is consistent with a suggestion originally made by Tukey (1981) and seconded by us (Ericksen and Kadane, 1985) and is an important design recommendation for the 1990 PEP.

8. CONCLUSION

We have considered a number of problems that arise when one attempts to apply the local estimation methodology to the problem of estimating local undercounts. None of the problems appears to be insurmountable, but the issue of extrapolating beyond the set of areas on which the regression equation is based is serious, and merits further attention. We feel that the difficulties occur mainly when local areas have values of independent variables that are extreme, and the problem is worse when these extreme values are outside those found in the areas on which the equation is calculated. This problem can be lessened through appropriate design, but we must also be sensitive to other problems of extrapolation occurring in unusual areas.

REFERENCES

Carter, G., and Rolph, J. (1974), "Empirical Bayes Methods Applied to Estimating Fire Alarm Probabilities," *Journal of the American Statistical Association*, **69**, 880–885.

Cowan, C. D., and Bettin, P. J. (1982), "Estimates and Missing Data Problems in the Postenumeration Program," Technical Report, U.S. Bureau of the Census.

Cuomo v. Baldrige, (1984), *Trial Transcript*, Federal Court for the Southern District of New York, Civil Action 80-4550.

DuMouchel, W., and Harris, J. (1983), "Bayes Methods for Combining the Results of Cancer Studies in Human and Other Species," *Journal of the American Statistical Association*, **78**, 297–305.

Ericksen, E. P. (1974), "A Regression Method for Estimating Populations of Local Areas," *Journal of the American Statistical Association*, **69**, 867–875.

Ericksen, E. P. and Kadane, J. B. (1985), "Estimating the Population in a Census Year (with comments and rejoinder)," *Journal of the American Statistical Association*, **80**, 98–131.

Fay, R. E., III, and Herriot, R. A. (1979), "Estimates of Income for Small Places: An Application of James-Stein Procedures to Census Data," *Journal of the American Statistical Association*, **74**, 269–277.

Freedman, D. A., and Navidi, W. C. (1986), "Regression Models for Adjusting the 1980 Census (with comments and rejoinder)," *Statistical Science*, **1**, 3–39.

James, W., and Stein, C. (1961), "Estimation with Quadratic Loss," *Proceedings of the Fourth Berkeley Symposium on Mathematical Statistics and Probability*, University of California Press, Berkeley, Vol. I, 361–379.

Lindley, D. V., and Smith, A. F. M. (1972), "Bayes Estimates for the Linear Model," *Journal of the Royal Statistical Society, Series B*, **34**, 1–19.

Morris, C. (1983), "Parametric Empirical Bayes Inference: Theory and Applications," *Journal of the American Statistical Association*, **78**, 47–55.

Siegel, J., Passel, J. S., and Robinson, J. G. (1980), "Preliminary Review of Existing Studies of the Number of Illegal Residents in the United States," paper presented to the Select Commission of Immigration and Refugee Policy.

Tukey, J. W. (1981), "Discussion of 'Issues in Adjusting for the 1980 Census Undercount' by Barbara Bailar and Nathan Keyfitz," paper presented at the Annual Meeting of the American Statistical Association, Detroit.

United States Bureau of the Census (1978), "The Current Population Survey: Design and Methodology," Technical Paper 40, U.S. Government Printing Office, Washington, DC.

United States Bureau of the Census (1982), "Coverage of the National Population in the 1980 Census by Age, Race, and Sex," *Current Population Reports, Series* P-23, No. 115, U.S. Government Printing Office, Washington, DC.

Warren, R., and Passel, J. S. (1983), "Estimates of Illegal Aliens from Mexico Counted in the 1980 United States Census," paper presented at the Annual Meeting of the Population Association of America, Pittsburgh.

Recent Developments in the Regression Method for Estimation of Population for Small Areas in Canada

R. B. P. Verma and K. G. Basavarajappa
Statistics Canada

ABSTRACT

The purpose of this paper is to present an overview of the recent developments in the application of the regression method in estimating population for small areas with special reference to Canada, and to compare the accuracy of this method with more traditional methods, such as that based on components of population change. It was found that neither the ratio-correlation nor the difference-correlation method uniformly or routinely outperformed the other. Hence, a multiple-model framework is adopted for evaluating the competing estimation techniques.

1. INTRODUCTION

Various methods exist for estimating the population for small areas. These include variations of the component method, vital rates method, proportional allocation, and regression method. None of these, however, is ideal in terms of accuracy, timeliness, and consistency. A multiple-model framework as an approach that takes account of strengths and advantages of each method has been found to be very satisfactory (Mandell and Tayman, 1982; Verma, Basavarajappa, and Bender, 1982b,

1982c, 1983, 1984). Among the regression-based methods, the ratio- or difference-correlation procedures are commonly used. These procedures employ a variety of symptomatic indicators that are readily available and that include births, deaths, school enrollments, hydro connections, family allowance recipients, drivers' licenses, tax filers, as predictors of population change. In recent years, the regression procedures have been widely used in different countries including Canada, the United States, Australia, and New Zealand (Statistics Canada, Catalogue #91-211, 1985a; Zitter and Cavanaugh, 1980; OPCS, 1982; Hughes and Choi, 1982). As the ratio- or difference-correlation procedures provide r^2 and standard errors, evaluation of these procedures and selection can be more objective than other techniques, such as component or vital rates methods. The ratio- or difference-correlation method uses the dependent or independent variables in the proportional form, which is a useful feature. It controls for differences in proportional sizes among small areas.

The purpose of this paper is to present an overview of the recent developments in the application of the regression method in estimating population for small areas with special reference to Canada and to compare the accuracy of this method with more traditional methods, such as that based on components of population change.

2. CHOICE OF ESTIMATION METHODS

In this section, descriptions of some of the estimation methods and their performance are given.

2.1. Regression Methods

The basic regression equation is

$$\Delta Y = \alpha + \beta_1 \Delta x_1 + \beta_2 \Delta x_2 + \cdots + \beta_n \Delta x_n + \epsilon \qquad (1)$$

where ΔY is the vector of changes in proportional values of the dependant variable (population) between times t and $t + r$; ΔX is the vector of changes in proportional values of the independent (symptomatic) variables; the β's are the regression coefficients; α is a constant; and ϵ is a vector of stochastic errors, such that

$$E(\epsilon) = 0 \quad \text{and} \quad E(\epsilon\epsilon') = \sigma^2$$

In ratio-correlation formulation,

$$\Delta Y = \frac{P_i(t+r)}{P(t+r)} \frac{P_i(t)}{P(t)}$$

and in difference-correlation formulation

$$\Delta Y = \frac{P_i(t+r)}{P(t+r)} - \frac{P_i(t)}{P(t)}$$

Where P_i refers to the population of the ith small unit (say) and P refers to that of the province or a larger unit so that $\Sigma P_i = P$. Similar definitions apply to ΔX's. Equation (1) is usually fitted by using the principle of least squares. Often, precise estimates of the regression coefficients (β) are obtained by considering the problems of multicollinearity and homogeneity of the variance term (ϵ).

Knowing the values of the regression coefficients, α and β, and of $P(t+r+1)$ and ΔX at time $t+r+1$, the population of a small area at time $t+r+1$, $P_i(t+r+1)$, can be calculated using six types of regression methods (ratio-correlation, weighted ratio-correlation, ridge-weighted ratio-correlation, difference correlation, weighted difference-correlation, and ridge-weighted difference correlation).

The weighted regression methods control heteroscedasticity. Ridge regression controls for multicollinearity.

The regression procedure is used for estimating the preliminary set of population totals (3–4 months after the reference date) of census divisions and census metropolitan areas. However, the updated or revised set of population totals (15–18 months after the reference date) are obtained by component procedure, which is subsequently described.

2.2. Component Method

Although the regression method gives timely estimates, that is, with a delay of 3–4 months, it provides only totals with no other details. However, the component method, which provides estimates after 15–18 months, provides information on the components of population change. The equation for the component method is

$$P_{t+r+1} = P_{t+1} + B - D + In - On + Ie - Oe$$

where P_{t+r+1} and P_{t+1} refer to population totals at time t and $t+1$, respectively; and r ranges from 1 to 5; B refers to births, D to deaths, In

to in-migrants, On to out-migrants, Ie to immigrants, and Oe to emigrants, all during time $t + 1$ to $t + r + 1$.

Births and deaths are obtained from Vital Registration Records, and in-migrants, out-migrants, and emigrants are estimated from Revenue Canada files. Immigration is obtained from Employment and Immigration Canada records. The component method using migration estimates from school enrollment data was also tested (Verma, Basavarajappa and Bender, 1982b).

2.3. Regression Nested

If the components and the base population (Census counts) are accurately known, the procedure is unassailable. The method is conceptually and operationally simple and precise. Furthermore, the procedure provides estimates of components of change for subprovincial areas that are consistent with those for the province. One consequence of this is that the regression estimates and the component estimates referring to a given point of time could be different and such divergence could increase with time. Hence, a method has been devised to tie these two sets of estimates together. The regression procedure is used to obtain change from one point of time to another, and this is added to the previous year's component estimate. The estimate thus obtained is termed "regression-nested" estimate.

2.4. Other Methods

The other methods tested included vital rate method, ratio method using the provincial administrative file, and proportional allocation method based on family allowance recipients.

2.5. A Brief Review of Performance

Schmitt and Crosetti and many others have claimed that the ratio-correlation method is one of the most accurate methods [using as the criterion, the mean absolute error (MAE)] (Goldberg, Rao, and Namboodiri, 1964; Swanson, 1978, NRC, 1980; Mandell and Tayman, 1982). Later, some researchers including Schmitt and Grier suggested that the difference-correlation method is an improvement over the ratio-correlation method (O'Hare, 1976). This was because the difference-correlation method produced constant mean, a lower mean square error

(MSE), higher intercorrelations among the variables, and a resulting higher squared value of the coefficient of multiple correlation (R^2). These features are often used to evaluate the fitting of a regression model and are considered desirable.

However, no consistent relationship between the higher R^2 and the MAE has been observed. The accuracy of population estimates produced by the regression method is highly dependent on the temporal stability of the regression coefficients. In this respect, a recent study has shown that the ratio-correlation method was more suitable than the difference-correlation method (Mandall and Tayman, 1982). The difference-correlation method produced higher multicollinearity than ratio-correlation. Due to this, the difference-correlation shows higher instability in the regression coefficients over time periods (Spar and Martin, 1979).

A review of both techniques has revealed that neither the ratio-correlation nor the difference-correlation method uniformly or routinely outperforms the other (O'Hare, 1980). This was also confirmed by Verma, Basavarajappa, and Bender (1982b). Thus, a choice of ratio- or difference-correlation is dependent on a thorough evaluation of the performance of the regression method based on past data.

In the light of the preceding findings, a multiple-model framework seems to be the most appropriate course for evaluating the competing estimation techniques.

3. EMPIRICAL EVALUATION OF METHODS

The evaluation measure used for testing the accuracy of the procedures and the symptomatic indicators was MAE. The MAE is defined as the mean of the percentage absolute difference between the estimated population total and the census population total. Those procedures and the symptomatic indicators that gave timely estimates with minimum MAE were chosen for producing estimates during the post-1981 period. The specifications of the regression procedures and the symptomatic indicators along the MAE are presented in Table 1 and the regression coefficients are presented in Table 2.

Except for Manitoba and British Columbia, the mean absolute error in all provinces is less than 2%. For CMAs, the estimates can be obtained in two ways: by ratio-correlation procedure or by aggregating the CD estimates, because several CDs or parts of CDs usually make up a CMA. Aggregating CD estimates to obtain CMA totals was found to provide more accurate totals than the ratio-correlation.

Table 1. Specifications of the Regression Method for Estimating the Population Totals for Census Divisions (CD) and Census Metropolitan Areas (CMA) in Each Province, Post-1981 Period[a]

Area/Province	Number of CD or CMA	Type[b]	Model Period	Symptomatic Indicator	Test 1981 MAE
Census Divisions	260				
Nfld.–P.E.I.	13	RC	1976–1981	F	1.27
N.S.	18	RC	1971–1976, 1976–1981	F	1.50
N.B.	15	RC	1976–1981	F	1.30
Quebec	76	RC	1976–1981	F	1.81
Ontario	53	RC	1976–1981	F	1.99
Manitoba	23	WDC	1971–1976, 1976–1981	F	3.13
Saskatchewan	18	DC	1976–1981	RP	0.62
Alberta	15	WRC	1976–1981	F, RP	1.89
B.C.	29	WDC	1971–1976, 1976–1981	F, Hydro	2.14
CMAs	24	RC	1976–1981	F	2.30
		Agg. CD[c]			1.07

[a] F: Family allowance recipients aged 1–14 years.
RP: Reference population obtained from Health Insurance Files.
MAE: Mean absolute error $= \dfrac{1}{N} \sum \dfrac{|\hat{P}_i - P_i|}{P_i} \times 100$
$\hat{P}_i$: Estimated population for Census Division.
P_i: Census population for Census Division.
N: Number of Census Divisions in a given Province.
RC: Ratio-correlation.
WDC: Weighted difference-correlation.
WRC: Weighted ratio-correlation.
CD: Difference-correlation.
CMAs: Census Metropolitan Areas.
[b] For a description of the types of regression methods, the readers are referred to the paper by O'Hare (1976).
[c] Agg. CD = CMA estimates obtained by aggregating the appropriate CD estimates. Calgary where the annual census count is done by the city is included.
Source: Statistics Canada (1985a).

Table 2. Coefficients of Regression, α and β, for Estimating the Population Totals of Census Divisions (CD) and Census Metropolitan Areas (CMA) in Each Province, Post-1981 Period[a]

Province	Model Type	Period	α	$\beta_1(F)$	$\beta_2(RP)$
Nfld.–P.E.I.	RC	1976–1981	.478	.514	—
N.S.	RC	1971–1976, 1976–1981	.467	.526	—
N.B.	RC	1976–1981	.503	.498	—
Quebec	RC	1976–1981	.385	.621	—
Ontario	RC	1976–1981	.256	.741	—
Manitoba	WDC	1971–1976, 1976–1981	.000	.609	—
Saskatchewan	DC	1976–1981	.000	—	1.086
Alberta	WRC	1976–1981	.088	.460	.476
B.C.	WDC	1971–1976, 1976–1981	.000	.376	.606
CMA	RC	1976–1981	.139	.862	—

[a] F: Family allowance recipients aged 1–14 years.
RP: Reference population.
Source: Table 1.

3.1. Accuracy of Census Division Estimates

A comparison of the accuracy of estimates obtained from regression and component procedures is presented in Table 3. It appears that each of the alternate methods (regression, regression-nested, and component) is superior to the methods used during the period 1976–1981. For Canada as a whole, among the alternate methods, regression-nested seems to be the most accurate with the lowest MAE, 1.7%. The regression-direct is observed to be less accurate than both the component and regression-nested procedures. This is true in all provinces except the province of Saskatchewan, where the regression estimates, which are based on the reference population from health insurance files as the symptomatic indicator, exhibit lower errors than either the component or the regression-nested procedures. In 5 out of 10 provinces the regression-nested is slightly more accurate than the component method.

In order to assess the relative accuracy of each of the alternate methods, the paired *t*-test was also carried out. For Canada as a whole, it was found that the differences were statistically significant at 1% level of significance between the estimates obtained from the regression-direct and component method. This is also true in Ontario and Quebec. In contrast, the differences were not statistically significant between the regression-nested and the component method, indicating that both these

Table 3. Evaluation of Population Estimates, June 1, 1981 (Mean Absolute Error %)

Province	Number of CD	Regression-Direct[a]	Regression-Nested	Component	Methods used During 1976–1981[b]
Nfld.–P.E.I.	13	1.36	0.67	1.00	2.6
N.S.	18	1.64	1.27	1.07	6.8
N.B.	15	1.59	1.05	1.06	3.3
Quebec	76	3.10	1.63	2.02	2.5
Ontario	53	2.17	1.26	1.21	1.5
Manitoba	23	3.33	2.57	2.58	4.4
Saskatchewan	18	1.43	1.96	2.10	2.0
Alberta	15	4.45	2.84	2.39	5.1
B.C.	29	2.45	2.50	2.39	9.2
Total	260	2.55	1.72	1.80	2.9

[a] The method uses as symptomatic indicators reference population for Saskatchewan and family allowance recipients for all other provinces. The model period for all provinces is 1971–1976. The procedures use weighted ratio-correlation for Alberta, weighted difference-correlation for British Columbia, and ratio-correlation for all other provinces.
[b] Methods used during 1976–1981: Component II: Prince Edward Island, Nova Scotia, New Brunswick, Manitoba, Alberta; and British Columbia; ratio method: Ontario and Saskatchewan; ratio-correlation: Newfoundland and Quebec. It should be noted that the symptomatic indicators were different. For a description of all these methods, reference may be made to Statistics Canada (1985a).
Source: Statistics Canada (1985a).

methods are statistically similar in terms of accuracy. Similar results were observed when the *t*-test was performed on the weighted MAE, which takes into account the size of the population.

3.2. Temporal Stability

In order to illustrate the temporal stability of regression, component, and regression-nested procedures, the index of dissimilarity was computed for the years 1977 through 1981 and is presented in Table 4. The index of dissimilarity is defined as one-half of the sum of absolute differences between two proportional distributions. Its value can range from 0 to 100 (see Table 5). It may be observed that both the disparities between the regression-direct and component estimates (A) and the regression-direct and regression-nested estimates (C) increase over time. However, the disparity between the regression-nested and component estimates fluctuates slightly and is found to be minimum. Thus, these two methods, in general, provide similar results during 1976–1981.

Table 4. Temporal Stability of (Regression-Direct, Regression-Nested, and Component Procedures) 1977–1981, Index of Dissimilarity[a]

Provinces		1977	1978	1979	1980	1981
Nfld.	A	0.17	0.33	0.41	0.34	0.51
	B	0.17	0.19	0.19	0.13	0.13
	C	0.00	0.17	0.34	0.35	0.41
P.E.I.	A	0.17	0.26	0.25	0.51	0.51
	B	0.17	0.08	0.19	0.02	0.24
	C	0.00	0.17	0.26	0.52	0.26
N.S.	A	0.29	0.53	0.60	0.63	0.64
	B	0.29	0.30	0.18	0.23	0.19
	C	0.00	0.53	0.38	0.45	0.70
N.B.	A	0.52	0.38	0.46	0.71	0.48
	B	0.52	0.48	0.44	0.52	0.37
	C	0.00	0.53	0.38	0.45	0.70
Quebec	A	1.02	0.64	0.81	0.99	1.13
	B	1.02	0.72	0.27	0.57	0.54
	C	0.00	1.05	0.66	0.80	0.98
Ontario	A	1.69	0.58	0.70	0.99	0.94
	B	1.69	1.75	0.31	0.49	0.56
	C	0.00	1.67	0.55	0.71	0.96
Manitoba	A	0.21	0.39	0.60	0.70	0.80
	B	0.21	0.26	0.26	0.21	0.19
	C	0.00	0.20	0.42	0.59	0.70
Saskatchewan	A	0.37	0.52	0.53	0.70	0.78
	B	0.37	0.18	0.26	0.25	0.18
	C	0.00	0.38	0.51	0.55	0.68
Alberta	A	0.45	0.45	0.57	0.89	1.18
	B	0.45	0.21	0.27	0.41	0.36
	C	0.00	0.44	0.43	0.56	0.86
B.C.	A	0.39	0.45	0.76	0.95	0.93
	B	0.39	0.32	0.41	0.23	0.29
	C	0.00	0.37	0.43	0.76	0.94

[a] Index of dissimilarity between estimates $\hat{P}_1$ and $\hat{P}_2$ for a province with n census divisions and total population P is given by

$$\frac{1}{2} \sum_{i=1}^{r} \frac{|\hat{P}_{1i} - \hat{P}_{2i}|}{P} \times 100$$

A: Index of dissimilarity between regression-direct and component estimates.
B: Index of dissimilarity between regression-nested and component estimates.
C: Index of dissimilarity between regression and regression-nested estimates.
Source: Statistics Canada (1985a).

Table 5.   Comparison of the Accuracy of the Regression Methods for the Model Periods 1971–1976 and 1976–1981[a]

	Regression		Model 1971–1976		Model 1976–1981
	Type	Indicator	Test 1976 MAE	Test 1981 MAE	Test 1981 MAE
Nfld.–P.E.I.	RC	F	1.6	1.4	1.3
N.S.	RC	F	1.8	2.0	1.6
N.B.	RC	V, F	2.0	1.0	0.9
Quebec	RC	V, F	1.4	2.3	1.8
Ontario	RC	V, F	2.0	2.5	2.1
Manitoba	RC	F	1.9	3.3	3.5
Saskatchewan	RC	RP	1.5	1.3	0.7
Alberta	RC	F	3.1	4.6	4.2
B.C.	WDC	F	3.1	4.0	2.3
Canada			1.96	2.54	2.04

[a] WDC: Weighted difference-correlation.
  RC: Ratio-correlation with ordinary least squares.
    F: Family allowance recipients aged 1–14 years.
    V: Vital events (births + deaths).
  RP: Reference population in Saskatchewan.
MAE: Mean absolute error (%).
Source: Statistics Canada (1985a).

The component and regression methods are independent, and so the results may be expected to diverge, whereas, the regression-nested and the component methods overlap and so the results may be expected to be similar.

The difference between the regression and the component estimates, which tends to be larger than the other two differences, is not expected to decrease because there are some theoretical weaknesses inherent in the regression method. For example, the assumption in the regression method that the vector of regression coefficients for symptomatic indicators is invariant from the immediately preceding intercensal period to the postcensal period is often questionable. In practice, this invariance may not hold good over time, both because of structural changes in the underlying relationships of the variables and also because of the improvement in the quality of the symptomatic indicators over time. Consequently, the model may fit well for the previous time period, but may predict poorly during the succeeding period.

3.3.   The Effects of Structural Changes

In order to examine the effects of structural changes during 1976–1981 on the accuracy of estimates the MAEs resulting from the equations of the model period 1971–1976 were compared with those resulting from the regression equations of the model period 1976–1981 (see Table 5). It may be seen that the 1981 mean errors resulting from the application of equations for two different time periods are quite comparable in all provinces except Saskatchewan and B.C., where the errors declined by nearly 50% (from 1.3% to 0.7%) and 40% (from 4.0% to 2.3%), respectively.

Due to structural changes, the relationship between the variations in symptomatic indicators (vital events and family allowance recipients) and variations in population have undergone changes during the period, 1976–1981. This is probably so for the provinces of Quebec, Manitoba, and Alberta. During the period 1976–1981, the characteristics of the people moving from the Maritime Provinces to the western provinces may have changed considerably. For example, as the family allowances are limited to the families with children eligible to receive family allowances, movement of single persons and families without children were not captured by the changes in the family allowance indicator. Due to this, the family allowance recipients as an important predictor of the population change in the regression model 1976–1981 failed to predict adequately. Thus, it is clear that the mean errors in 1981 resulting from models of both time periods, 1971–1976 and 1976–1981 were high, because of structural changes.

A part of the difference in the mean errors between 1976 and 1981 is also due to changes in the quality of family allowance data. The numbers of family allowance recipients are produced at the census division level by converting postal codes to standard geographic codes. In 1976, the conversion file had problems with missing the overlapping postal codes. In particular, the percentage of missing codes in family allowance files in 1976 was high in Maritime Provinces and Ontario. But, by 1981, the magnitude of the problem of missing postal codes in the FA files had declined in all provinces. Thus, the change in the quality of the family allowance indicator between the years 1976 and 1981 may have also affected the quality of the regression coefficients during the period 1976–1981.

3.4. Accuracy of CMA Estimates

Table 6 presents the MAEs of 1981 population estimates for CMAs. They are produced by the regression-direct, regression-nested, and component methods based on CMAs as units and also CMA totals obtained by aggregating census division estimates. It may be seen that for each

Table 6.   Evaluation of 1981 Population Estimates for CMAs, MAE (%)

Method	Based on Ratio-Correlation	Agg. CDs[b]
Regression	2.25[a]	1.77
Regression-Nested	2.21	1.03
Component	1.47	1.20

[a] Regression (F), 1971–1976. F: family allowance recipients aged 1–14 years.
[b] Agg. CDs = Aggregating census divisions estimates.
Source: Demography Division, Statistics Canada, (1985b).

method, the errors in estimates obtained by aggregating census division totals are considerably lower than those based on ratio-correlation. The larger errors in CMA estimates based on ratio-correlation may be due to coverage problems contained in the postal code translation files. Regression-nested procedure based on aggregating CD totals seems to provide the least error. Paired comparison of the three methods may be made by considering the indices of dissimilarity, which are given below:

Comparison	Estimates Based on Ratio-Correlation	Agg. CDs
Nested versus component	0.98%	0.48%
Regression-direct versus component	1.15%	0.86%
Regression-direct versus nested	1.09%	0.54%

The indices of dissimilarity show that the differences among the estimates are small and that the regression-nested and the component methods provide very similar results.

3.5.   Remarks on Accuracy, Consistency and Timeliness of Estimates

In terms of accuracy and consistency with respect to sources of input data and methods used for estimating the population of different geographic units (provinces and territories, CDs and CMAs), the component method appears to be the best. In addition, the component method provides more detailed and consistent information on components of population change, for example, consistent set of internal migration figures classified by streams (in and out) and by broad age groups and sex for the province and its subprovincial areas. However, this method does not provide timely estimates. The delay is usually about 15–18 months. The regression method using family allowance recipients and/or other variables as symptomatic indicators, on the other hand, provides estimates that are timely (i.e., with a delay of 3–4 months) but slightly less accurate. The regression-nested procedure provides estimates that are timely and are of accuracy roughly comparable to those obtained by component method.

4.   SUMMARY AND DISCUSSION

The regression procedure can be used for providing timely estimates for small areas. The accuracy of the population estimates depends on the stability of regression coefficients. The effect of instability can be reduced markedly by using the regression procedure to provide only annual changes in total population. The annual changes thus obtained could then be combined with an independently prepared component estimate that uses for its base the census total and annual changes derived from administrative or other sources. The estimate thus obtained has been termed the "regression-nested" estimate, and this was found to be roughly of the same accuracy as the component estimate.

As the error in estimate tends to be associated inversely with the size of the locality, it is sometimes suggested that for smaller areas, one may be better off using the previous census total than the regression estimate, or, alternatively, using a weighted average where preponderant weight is given to the census total and a smaller weight to the regression estimate. The regression-nested procedure automatically incorporates such suggestions as the regression-nested estimate, is derived by adding the estimated annual change in population obtained by regression to the previous years component estimate. Consequently, the regression-nested estimate may be composed of 80–90% of the census total and 10–20% of the regression estimate, depending on the rate of growth of the area during the intercensal period. Furthermore, it seems desirable to use the regression procedure only to provide timely preliminary estimates and to anchor the updating or revision of these estimates firmly in component procedure as long as the components of growth can be estimated from administrative or other sources.

It should be noted that even when the tests of regression procedures based on comparisons with the past census totals showed excellent results, there is no assurance that the procedure will work as satisfactorily for postcensal years as it did for past periods.

The selection of most appropriate symptomatic indicators is very important. In Canada, apart from the family allowance recipients aged 1–14 years, which is currently being used, others—hydro-connections, drivers' licences, telephone connections, and population counts from Health Insurance files—are being explored. Among all these, health insurance files have the highest potential. They are universal in coverage and thus come close to the "continuous population registers" maintained in some European countries. They have been partially developed in all provinces except Ontario. At some future date if Ontario comes into the fold, the files would have the potentiality to provide the interprovincial migration figures and the total counts themselves.

ACKNOWLEDGMENT

We are thankful to Dr. M.P. Singh of Statistics Canada for many helpful comments on an earlier draft.

REFERENCES

Almond, M. M. (1982), *An Inventory of Sources of Canadian Migration Data*, Statistics Canada, Ottawa, Mimeographed, pp. 1–88.

Bender, R. K., and Verma, R. B. P. (1983), "Translation for Converting Demographic Data Between Overlapping Sub-Provincial Areas in Canada, 1983," *Proceedings of the Social Statistical Section*, American Statistical Association, Washington, DC, pp. 518–521.

Cliff, A. D., and Ord, J. K. (1981), *Spatial Processes Models and Applications*, Pion Limited, London.

Cliff, A. D., and Ord, J. F. (1975), "Model Building and the Analysis of Spatial Pattern in Human Geography," *Journal of the Royal Statistical Society, Sec. B*, **37**, 297–348.

Dominion Bureau of Statistics (1967), *Population Estimates for Counties and Census Divisions, Catalogue No. 91-206*, The Queen's Printer, Ottawa.

Goldberg, D., and Balakrishnan, T.R. (1960). "A Partial Evaluation of Four Estimation Techniques," paper presented at the Annual Meeting of the Social Statistics Section, American Statistical Association.

Goldberg, D., Rao, V.R., and Namboodiri, N.R. (1964), "A Test of the Accuracy of the Ratio Correlation Population Estimates," *Land Economics*, **40**, 100–102.

Government of British Columbia Central Statistics Bureau (1980). "British Columbia School District Population Estimates, and also, "British Columbia Municipal Population Estimates," Ministry of Industry and Small Business Development, Government of British Columbia, Victoria, unpublished report.

Hughes, P. J., and Choi, C. Y. (1982), *Regression Techniques for LGA Population Estimation*, Demography and Social Branch Australian Bureau of Statistics, Canberra, pp. 1–24.

Johnson, T. (1963), *Econometric Methods*, McGraw-Hill, New York, p. 219.

Katzoff, M. J., and Malec, D. (1983), "Application of Some Common Spatial Models to Population Data," *Proceedings of the Social Statistics Section*, American Statistical Association, Washington, DC, pp. 522–524.

Mandell, M., and Tayman, J. (1982), "Measuring Temporal Stability in the Regression Models of Population Estimation," *Demography*, **19**(1), 135–146.

National Research Council (NRC) (1980), *Estimating Population and Income of Small Areas*, National Academy Press, Washington, DC, pp. 1–247.

Norris, D. A., Britton, M., and Verma, R. (1982), "The Use of Administrative Records for Estimating Migration and Population," Presented at the Annual Meeting of the American Statistical Association, Cincinnati, OH, August 16–19.

Norris, D. A., and Standish, L. D. (1983) "A Technical Report on the Development of Migration Data from Taxation Records," Statistics Canada, Ottawa (draft).

Ord, K. (1975), "Estimation Method for Models of Spatial Interaction," *Journal of the American Statistical Association*, **70**, 120–126.

Office of Population Censuses and Surveys (OPCS) (1980), "A Comparison of the Registrar General's Annual Population Estimates for England and Wales with the results of the 1981 Census," Occasional Paper 29. Population Statistics Division, OPCS.

O'Hare, W. (1976), "Report on a Multiple Regression Method for Making Population Estimates," *Demography*, **13**, 369–380.

O'Hare, W. (1980), "A Note on the Use of Regression Methods in Population Estimates," *Demography*, **7**, 87–92.

Raby, R., and Parent, P. (1982), *Postcensal Emigration Estimates 1980–1982*, Statistics Canada, Ottawa, Mimeographed, pp. 1–29.

Romaniuc, A., Raby, R., and Parent, P. (1982), "*The Choice of Methods for Estimating Interprovincial Migration for the Post-1981 Period*, Ottawa, Statistics Canada, Ottawa, Mimeographed, pp. 1–43.

Spar, M., and Martin, J. (1979), "Refinements to Regression-Based Estimates of Postcensal Population Characteristics," *Review of Public Data Use*, **7**(5/6).

Statistics Canada, Annual. *Estimates of Population for Census Divisions*, Cat. 91-206, Ministry of Supply and Services, Government of Canada, Ottawa.

Statistics Canada, Annual, *Estimates of Population for Census Metropolitan Areas of Canada*, Cat. 91-207, Ministry of Supply and Services, Government of Canada, Ottawa.

Statistics Canada (1985a), *Postcensal Annual Estimates of Population for Census Divisions and Census Metropolitan Areas, June 1, 1982 and 1983 (Regression Method)*, Catalogue 91-211, Ministry of Supply and Services, Government of Canada, Ottawa.

Statistics Canada, (1985b), *Postcensal Annual Estimates of Population for Census Divisions and Census Metropolitan Areas, June 1, 1982 (Component Method)*, Ministry of Supply and Services, Government of Canada, Ottawa.

Swanson, D. A., and Tedrow, L. M. (1984), "Improving the Measurement of Temporal Change in Regression Models Used for County Population Estimates," *Demography*, **21**(1) 373–381.

Swanson, D. (1978), "An Evaluation of Ratio and Difference Regression Methods for Estimating Small, Highly Concentrated Population: The Case of Ethnic Groups," *Review of Public Data Use*, **6**, 18–27.

Verma, R. B. P., and Basavarajappa, K. G. (1982a), "A Sub-provincial Estimation of the Population in Canada: A Review of Estimation Methods and Prospects for Development," Presented at the Population and Family Planning Session, American Public Health Association Meeting, Montreal, November, pp. 1–32.

Verma, R. B. P., Basavarajappa, K. G., and Bender, R. (1982b), *New Approaches to Methods of Estimating the Population of Census Divisions*, Statistics Canada, Ottawa, Mimeographed, pp. 1–77.

Verma, R. B. P., Basavarajappa, K. G., and Bender, R. (1982c), *New Approaches to Methods of Estimating the Population of Census Metropolitan Areas*, Statistics Canada, Ottawa, Mimeographed, pp. 1–39.

Verma, R. B. P., Basavarajappa, K. G., and Bender, R. (1983), "The Regression Estimates of Population for Sub-provincial areas in Canada, *Survey Methodology*, 9(2) 219–240.

Verma, R. B. P., Basavarajappa, K. G., and Bender, R. (1984a), "Estimation of Local Area Population: An International Comparison, 1984," *Proceedings of the Social Statistics Section*, American Statistical Association, Washington, DC, pp. 324–329.

Verma, R. B. P., Basavarajappa, K. G. and Bender, R. (1984b), "Generalized System for Evaluation and Production of Total Population estimates for Sub-provincial Areas," presented at the Annual Meeting of the Population Association of America, Minneapolis, MN, May 3–5.

Zitter, M. and Cavanagh, F. J. (1980), "Postcensal Estimates of Population," An unpublished paper presented at the Annual Meeting of the American Association for the Advancement of Science, Session on the 1980 Census, San Francisco, CA, January 1980.

Regression Methods and Performance Criteria for Small Area Population Estimation

P. McCullagh
University of Chicago

J. V. Zidek
University of British Columbia

ABSTRACT

In this chapter a regression model is developed for postcensal estimation of the population sizes of small areas. The approach is nonstochastic. It is assumed that current population sizes have been determined by a function of those obtained at the last census together with the associated values of certain symptomatic variables. As well, the current values of these variables are assumed to be inhand. Natural properties of such a rule are shown to imply a specific, log-linear form for this function. Existing models are shown to be approximations to the result. An objective function for evaluating arbitrary estimation procedures or fitting regression models is derived on the assumption that revenue allocation is the objective of the estimation program. By appealing to an equilibrium theory for group decision processes, it is shown that under certain conditions the appropriate criterion is given by the Kullback–Leibler discrimination function.

1. INTRODUCTION

Government and private agencies depend on estimates of the populations of small areas for a variety of purposes such as revenue allocation and planning. In the United States, for example, annual, postcensal estimates are prepared under the Federal–State Cooperative Program (FSCP) for something like 39,000 municipalities, counties, and so on (Kitagawa et al., 1980).

The great scale on which this activity is carried out makes the need for accurate, simply applied procedures acute. A variety of methods exist (Purcell and Kish, 1979; Kitagawa et al., 1980; Zidek, 1982). None is simpler and more adaptable than those that use regression models for making these estimates (Schmitt and Crosetti, 1954; Morrison and Relles, 1975; O'Hare, 1976). These models turn easy-to-measure or readily available quantities (symptomatic variables) into estimates of population sizes that are expensive to measure. The coefficients obtained by fitting these models using two successive censuses and the measured values of all variables, one set per subarea, (implicitly) account for migration, demographic trends, and so on, so that the latter are not needed in the preparation of the estimates as they would be in, say, administrative records methods. Different symptomatic variables may be used in different subareas, depending on what data are readily available in each. And comparative empirical studies in the previously cited works, show that these methods can be very effective, surprisingly so since the regression methodology is misappropriated in the sense that there is no conceivable experiment that generates these variable values at random as in the conventional case. One such method, that using the ratio-correlation model, is among the three that constitute the overall strategy used in the FSCP (Kitagawa et al., 1980).

In spite of the relatively high precision of regression-based methods and their simplicity and adaptability, there have been surprisingly few models proposed, and these have not been refined very much. Rosenberg (1968) points out the need to stratify local areas by dichotomies, first of urban versus rural and second of rapid versus slow growth. His proposal seems to have been ignored. Namboodiri (1972) argues persuasively against the needed temporal stationarity of regression models and gives an analysis that may well show the ill-effects of multicollinearity among the symptomatic variables. A systematic residual analysis has not been published and little seems to be known about the possibly serious negative impact of influential observations in this context. [See Belsey, Kuh, and Welsch (1980) for a general discussion of this problem.]

Existing regression models have been proposed on an ad hoc basis. The major result of Section 3 is a new model, which is obtained by what might be called an axiomatic approach. Stochastic and approximation errors aside, it is supposed that population sizes of small areas are assigned, hypothetically, by some unspecified function of the symptomatic variables and previous counts. The most reasonable requirements of such a rule are specified and then shown to be equivalent to a log-linear model similar but not identical to that of Morrison and Relles (1976). The model has yet to be assessed empirically.

Both the fitting of regression models and the assessment of estimation methods requires a criterion function that accumulates in some sensible way the errors made over all subareas. No particular choice has yet been indicated, and various alternatives such as average relative absolute error are used, sometimes several in the same study. None of these criteria seem directly related to a primary objective of the program, namely, the allocation of revenue. In Section 2 a criterion function is derived from first principles. This function takes account of the need to choose an allocation scheme which would be a jointly acceptable compromise to all members of the community, provided their desires obey certain weak constraints. The derivation is based on the theory of Nash (1950).

2. CRITERIA

Evaluating the performance of an estimation methodology requires answers to two fundamental questions. Against what are the answers it produces to be compared and by what criterion is the comparison to be made? This section is addressed to the second of these questions. We have no alternative to propose to the practical answer that is commonly given to the first question, namely, the corresponding census counts. The latter typically underestimate the true counts by something like 2% (Hauser, 1981) and therefore seem somewhat unsatisfactory.

Kitagawa et al. (1980) describe various estimation performance criteria in terms of P_i the "actual" and $\hat{P}_i$ the estimated population sizes for subregion $i = 1, \ldots, n$. The former would be the last available census counts corrected, possibly, as Spencer suggests (Kitagawa et al., 1980, Appendix 1) for undercoverage. These criteria include average error, average relative error, number of extremely large relative errors, and bias. The first two of these are obtained by putting $a = 1$, $b = 0$ and $a = 1$, $b = 1$, respectively, in the general index $\Sigma |P_i - \hat{P}_i|^a / P_i^b$.

While Kitagawa et al. (1980) refer to the need to take account of the purposes of the estimation program, no criterion has been given that does so, other than that which is derived below. Its derivation relies on Nash's theory of bargaining as it might be applied in the present situation (Nash 1950). Let P and A denote, respectively, the population size and amount (of revenue, say) to be allocated. Let $A = (A_1, \ldots, A_n)$ be a feasible allocation of A among the n subregions and $u_{ij}(A)$, $j = 1, \ldots, P_i$, $i =$

$1, \ldots, n$, represent the gain-in-value (utility) to individual j in subregion i that would result from this allocation scheme. Certain weak assumptions imply that any equilibrium solution must maximize an objective function, which will now be described.

Before doing so, it should be pointed out that Nash's theory admits as potential solutions not only the allocations A themselves but all random-ized mixtures of the A's. Deadlocks can therefore be broken by tossing a coin, as it were. The domains of the u_{ij}'s are extended to this more general solution set by invoking the expected utility hypothesis. However, the feasible solutions, randomized and nonrandomized alike, are required to satisfy $u_{ij} \geq 0$. In this way, the Nash theory ensures that no individual can be made to suffer a net expected loss of utility as a result of the proposed allocation.

In agreement with practice, we will restrict ourselves to nonrandom-ized allocations and define as optimal any that are feasible and maximize the so-called Nash product given by

$$NP(A) = \prod_i \prod_j u_{ij}^{1/P}(A) . \tag{1}$$

If the u_{ij}'s are unknown, as would be the case in the situation under consideration, an approximation to equation (1) becomes necessary. If the A_i's are moderate, it is reasonable that $u_{ij}(A)$ would equal, approxi-mately, A_i/P_i for all i and j. This approximation is supported by the assumptions that the ith subregion's allocation is evenly distributed among its constituents, that they receive no benefit from the allocations made to other subregions, and that the utilities are linear. This approxim-ation yields

$$NP(A) = \prod (A_i/P_i)^{p_i}. \tag{2}$$

where $p_i = P_i/P$ is the proportion of the population in subregion i for all i.

Equation (2) may be reduced further by letting $A_i = a_i A$, $i = 1, \ldots, n$. Then

$$NP(A) = (A/P) \exp[-I(p, a)] \tag{3}$$

where $I(p, a) = \Sigma\, p_i \log(p_i/a_i) \geq 0$ is the Kullback–Leibler "distance" between a and p.

To maximize NP is to minimize I, that is, to choose $a = p$. However, in practice p would not be known. It would seem natural then to choose for a the best available estimate, say, $\hat{p}$ of p.

The approximate *NP*-criterion given in equation (3) can be obtained by entirely different reasoning. Suppose a random sample of the region's current population is drawn with replacement and each individual so obtained is classified by subregion. Let p_i be the observed sample fraction of individuals from subregion i. Then the function of *a* given in equation (3) is the likelihood function for these data. It would be maximized to find the optimal estimate among allowable choices of *a* in order to find the maximum likelihood estimate of the true subregion population proportions. This would be *a* = *p* unless *a* were constrained.

This sampling-theoretical point of view suggests a natural alternative to the criterion given in equation (3). If the hypothetical sample were large and the a_i's were the "true" regional proportions, then the consistency of the (unconstrained) maximum likelihood estimator, *p*, would yield the approximation $\log(a_i/p_i) \simeq (a_i/p_i - 1) - (a_i/p_i - 1)^2/2$. This, in turn, would yield

$$NP(A) = (A/P) \exp\left(-\tfrac{1}{2} \sum (a_i - p_i)^2/p_i \right) \tag{4}$$

Equation (4) suggests the minimum chi-squared criterion for choosing the $\{a_i\}$, namely, minimize

$$\chi^2 = \sum (a_i - p_i)^2/p_i \tag{5}$$

Again, if *a* were unconstrained, *a* = *p* would be optimal.

This last criterion among others receives special attention in Kitagawa et al. (1980). It represents, according to these authors, a compromise between $\Sigma (a_i - p_i)^2$ and $\Sigma |a_i - p_i|/p_i$. The first would be unduly sensitive to large individual, subregional misallocations and the second, to misallocations in small areas.

3. REGRESSION METHODS

Such methods yield postcensal estimates, $\hat{P}_{2i}$, of current (time $t = 2$, say) population sizes, P_{2i}, for subregions. They rely on models that involve coefficients which must be fitted, the observed current values of symptomatic variables, $S_{2i} = (S_{2i1}, \ldots, S_{2ip})$, and their corresponding values at time $t = 1$, S_{1i}, when population sizes, P_{1i}, are available for all regions, $i = 1, \ldots, n$.

To fit the coefficients, a criterion is chosen, usually least squares but possibly that given by I or χ^2 in equations (3) or (5), respectively. The times $t = 1$ and 2 are taken to be successive census years so that the $\{P_{ij}\}$, $\{S_{ji}\}$ $j = 1, 2, i = 1, \ldots, n$ are observable. Then p_i is taken to be the observed value of the "dependent" variable, $P_{2i}/\Sigma\, P_{2i}$, $i = 1, \ldots, n$, while a_i is supplied by the regression model, albeit with as yet unspecified coefficients. The coefficients are then chosen to give the criterion-functional its least possible value.

Erickson (1973, 1974) proposed an interesting variation of this scheme, which might be called "sampling-regression." The time $t = 2$ is the present, and the model is fitted to the results of a census carried out on a subsample of subregions. This approach takes account of the inevitable temporal nonstationarity of any regression model by "tuning" its coefficients to the present. The sampling scheme that was introduced in Section 2 to provide an interpretation of the NP criterion would provide a (no doubt inferior) alternative to Erickson's plan. A bonus of the sampling-regression approach, which is pointed out by Purcell and Kish (1979), is that, unlike the conventional approach, it carries with it a basis for inference.

It remains to choose a suitable model. As will now be shown, simple intuitive requirements lead quite easily to particular models.

For expository simplicity it will be assumed that $p = 1$ so that the symptomatic variables at times $t = 1$ and $t = 2$ become S_{1i} and S_{2i}, respectively. The "dot" will as usual represent summation over any subscript it replaces. Thus $P_{2.}$ would denote the population of the entire region at time $t = 2$.

Let $\hat{P}_2 = T(S_2, S_1, P_1)$ represent an unspecified postcensal regression estimation model. Here $T = (T_1, \ldots, T_n)^T$ is the vector of subregional estimation models, while $S_{ji} = (S_{ji}, \ldots, S_{jn})^T$, $j = 1, 2$, and $P_1 = (P_{11}, \ldots, P_{1n})^T$ are the corresponding vectors of symptomatic variables and population sizes at time $t = 1$. Observe that, in general, T_i is a function of S_2, S_1, and P_1 for all i.

Suppose, hypothetically, that subregional population sizes were to be assigned without error by means of the model, T previously introduced. Various reasonable requirements of such a rule suggest themselves and lead, as is shown below, to explicit forms for T. These may be taken as first approximations to the actual population sizes and so are used to obtain estimates. The precision of such estimates and hence the value of the models suggested by the approach outlined previously would need to be ascertained by empirical study.

Possible requirements for T are presented and discussed. Their implications are given below.

Regional population sizes may be estimated with a relatively high precision compared to those of its subregions. So they may be regarded as

known. The following requirement is therefore considered as fundamental:

$$\text{CTT } (\textit{controlled-to-total}): \quad \hat{P}_{2\cdot} = P_{2\cdot}.$$

It implies that we may regard P_2 and P_1 as vectors of proportions, each set
summing to 1, rather than as vectors of population counts whenever it is
expedient to do so.

Another important condition is

PI (*Permutation Invariance*):   *For all $n \times n$ permutation matrices Φ*
$$T(\Phi S_2, \ \Phi S_1, \ \Phi P_1) = \Phi T(S_2, S_1, P_1)$$

This condition simply ensures that the order in which the subregions are
listed is irrelevant. It is not the same as spatial homogeneity because the
symptomatic variables, S_1 and S_2, may contain information pertaining to
the geography of the subregions.

The following invariance requirements ensure that the model has good
robustness properties. They make use of the fact that the scale on which
the symptomatic variables are measured is usually arbitrary. Furthermore, if the symptomatic variables are counts computed locally (e.g.,
births, marriages, deaths), there may be a degree of underreporting that
varies from one region to another. To some extent these effects are
eliminated if the following invariance conditions are satisfied:

RI (*Regional Invariance*):   *For all positive diagonal matrices D*
$$T(DS_2, DS_1, P_1) = T(S_2, S_1, P_1)$$

TI (*Temporal Invariance*):   *For all positive scalars $a_1, a_2 > 0$*
$$T(a_2 S_2, a_1 S_1, P_1) = T(S_2, S_1, P_1)$$

The next condition embraces a natural equivariance requirement. It
derives from the recognition that the estimated growth rate in each
region, i, $\hat{P}_{2i}/P_{1i}$ would not change even if P_1's coordinates were replaced
by their corresponding densities, say, per hectare or per square kilome-
ter, for example. This suggests the condition

$$T_i(S_2, S_1, P_1)/(P_{1i}) = T_i(S_2, S_1, DP_1)/(d_i P_{1i}), \quad i = 1, \ldots, n,$$

for any positive $D = \text{diag}\{d_1, \ldots, d_n\}$.

This is equivalent to $T(S_2, S_1, DP_1) = DT(S_2, S_1, P_1)$, that is,
$T_i(S_2, S_1, P_1) = P_{1i} H_i(S_2, S_1)$, $i = 1, \ldots, n$ for some function H. However, this condition is inconsistent with CTT, which is considered more

fundamental. And it is unnessarily strong for it is possible, as our analysis will show, to derive a sufficiently small class of potential models under the weaker requirement that $q_{ij}(S_2, S_1, P_1) \overset{\triangle}{=} (\hat{P}_{2i}/P_{1i}) \div (\hat{P}_{2j}/P_{1j})$, the relative growth rates of regions i and j, $i, j = 1, \ldots, n$, be invariant in the sense described above. This condition is $g_{ij}(S_2, S_1, DP_1) = g_{ij}(S_2, S_1, P_1)$, $i, j = 1, \ldots, n$, for all positive diagonal matrices D, which is equivalent to

RGRI (*Relative Growth Rate Invariance*). For $i, j = 1, \ldots, n$ and all positive $D = \mathrm{diag}\{d_1, \ldots, d_n\}$,

$$G_{ij}(S_2, S_1, DP_1) = G_{ij}(S_2, S_1, P_1)$$

where

$$G_{ij}(x, y, Dz) = [T_i(x, y, Dz)/d_i][T_i(x, y, Dz)/d_j]^{-1}$$

for all x, y, and z in T's domain.

Instead of regarding, as we may because of CTT, P_1 and P_2 as vectors of proportions, it is more convenient to deal with the $(n-1)$-vector of ratios, P_1^*, with elements P_{il}/P_{in}, $i = 1, \ldots, n-1$. Similar changes in T yields $\hat{P}_2^* = T^*(S_2, S_1, P_1^*)$. The advantage of P_2^* over P_2, the vector of proportions, is that the elements of P_2^* are unrestricted even when CTT is imposed.

Conditions RI and TI are equivalent to $T^*(S_2, S_1, P_1^*) = T^*[(S_2/S_1), 1, P_1^*]$ and RGRI to $T_i^*(S_2, S_1, P_1^*) = P_i^* T_i^*(S_2, S_1, 1)$, $i = 1, \ldots, n$. So RI, TI, and RGRI combined are equivalent to

$$T_i^*(S_2, S_1, P_1^*) = P_i^* H_i[(S_2/S_1)^*] \tag{6}$$

for some function H, $i = 1, \ldots, n$.

We regard CTT, PI, RI, TI, and RGRI as important: the first two would seem to be essential, but it is possible to think of conditions under which the invariance requirements are not so compelling. The following conditions, although plausible, seem less important. They, or conditions like them, are needed to reduce the general model given in equation (6) to a more explicit, applicable form:

TC (*Temporal Coherence*): *Given an additional time, $t = 0$,*
$$T(S_2, S_0, P_0) = T(S_2, S_1, \hat{P}_1),$$

where

$$\hat{P}_1 = T(S_1, S_0, P_0)$$

TR (*Time Reversibility*): For all S_1, S_2, P_1,
$$P_1 = T(S_1, S_2, \hat{P}_2),$$

where

$$\hat{P}_2 = T(S_2, S_1, P_1)$$

The intuitive basis for TC is clear, and it nearly implies TR since the latter becomes the former when the past is reflected into the future provided $T(S_0, S_1, P_1)$ is given the value P_0, the subregions' known population sizes when the symptomatic variable S_0 reassumes its original (time $t = 0$) value.

These conditions imply an easily derived explicit form for the function H_i of equation (6). The solution, $f(y) = ay$ of Euler's functional equation $f(x + y) = f(x) + f(y)$, f continuous, is used. The solution obtains even if the requirement $x + y \le K$ is imposed, $0 \le x, y \le K$. The result is

$$H_i(y) = \prod_{j=1}^{n} y_j^{\alpha_{ij}} \tag{7}$$

for certain constants, $-\infty < \alpha_{ij} < \infty$.

If condition PI is imposed, a straightforward argument which is omitted for brevity shows that $\alpha_{ij} = \alpha$ or 0 according to whether $i = j$ or $i \ne j$ for some constant α. The model implied by all of the conditions given above is, then, in summary,

$$\hat{P}_{2i} = P_2 . P_{1i} R_i^\alpha \left(\sum_{k=1}^{n} P_{1k} R_k^\alpha \right)^{-1} \tag{8}$$

where $R_i = S_{2i}/S_{1i}$. Of course, P_{1i} and R_i can be replaced by P_{2i}^* and R_i^* in equation (8) without changing the result. Alternatively, they can be replaced by their corresponding "shares," P_{1j}/P_1. and R_i/R., respectively. In the case of more than one symptomatic variable, obvious extensions of the conditions stated previously lead to a generalization of the model given in equation (8):

$$\hat{P}_{2i} = P_2 . P_{1i} \prod_{j=1}^{p} R_{ij}^{\alpha_j} \left(\sum_{k=1}^{n} P_{1k} \prod_{j=1}^{p} R_{kj}^{\alpha_j} \right)^{-1} \tag{9}$$

where R_{ij}, the counterpart of R_i, is the growth rate of symptomatic variable j, $j = 1, \ldots, p$.

An alternative form of the previous model is obtained by taking logarithms. The result is a variant of that suggested by Morrison and Relles (1976):

$$\log(\hat{P}_{2i}/P_{1i}) = \prod_{j=1f}^{p} \alpha_j \log R_{ij} \tag{10}$$

The basis for their choice is not given. The resulting estimate $\hat{P}_2$. would not necessarily satisfy CTT and would therefore need to be "controlled" after the parameters α_j, $j = 1, \ldots, p$, where fitted. The resulting estimates would, in general, differ from those obtained by applying the model given in equation (9). The model given in equation (9) has not undergone a comparative study. That in equation (10) has been proposed on theoretical grounds and tested by Swanson and Tedrow (1984).

Approximations to these models may be derived using the approximation $\log x \simeq x - 1$, which is accurate if $x \simeq 1$. So, in stable subregions i, that is, those undergoing slow changes in population size, equation (10), for example, would yield the approximate model

$$\hat{P}_{2i}/P_{1i} = \alpha_0 + \sum_{j=1}^{p} \alpha_j R_{ij} \tag{11}$$

This model, or that obtained by replacing all its components by their shares, is called the ratio-correlation model. It is the most commonly used of the various alternatives and is the regression component of the composite methodology used in the U.S. FSCP (Kitagawa et al., 1980). Empirical studies presented in this last-cited work show that this model yields very good results in exactly those subregions where the above approximation would be accurate. The Swanson–Tedrow study previously cited indicates that the Morrison–Relles model is better but not markedly better than the ratio-correlation method.

If in the model of equation (10) the P's are replaced by their shares and logarithms are taken, an approximate model is obtained:

$$\hat{p}_{2i} - p_{1i} = \alpha_0 + \sum_{j=1}^{p} \alpha_j (r_{2ij} - r_{1ij}) \tag{12}$$

where $p_{ki} = P_{ki}/P_k$. and $r_{kij} = S_{kij}/S_{k\cdot j}$, $k = 1, 2$, $i = 1, \ldots, n$, and $j = 1, \ldots, p$. This model would be expected to be appropriate for subregions whose population sizes comprise a relatively large fraction of that of the region. This so-called difference-correlation model is proposed by O'Hare (1976), who provides empirical results which show that this model is marginally superior to its ratio counterpart. However, under this model, the data may well be heteroscedastic (Dr. D. Herman, personal communication) so its value remains uncertain.

4. CONCLUDING REMARKS

This paper presents a new regression model [equation (9)] for estimating the population sizes of small areas. Its value remains to be determined by empirical study. However, the argument of Section 3 makes it a natural choice. The ratio-correlation model in current use in the U.S. FSCP would approximate the proposed model in subregions of stable population size for reasons given in Section 3.

In fitting and empirically evaluating a model such as derived in Section 3, a criterion of "objective function" must be specified. That obtained in Section 2 [equation (3)] is novel in this context. It is, approximately, the Nash criterion (Nash, 1950) for determining the joint value to a group (society) of any proposed group action (revenue allocation scheme). It implies that proportional allocation on the basis of population size is optimal. Thus the value of an allocation scheme will depend on how well the best available estimates of subregional population proportions approximate the exact values.

The proposed criterion is the product of an attempt to relate the performance of an estimation method to one of the primary objectives of the estimation program, albeit under the oversimplification of linear utility functions. While this criterion may be used in evaluating any estimation procedure through the estimates of proportions, and, hence, the approximate allocation scheme it produces, its form is, fortuitously, particularly appropriate for the regression model developed in Section 3.

An anonymous referee has argued, "in general it is a good idea to calculate various measures of performance for a given method. If a method performs well according to most measures of performance, the confidence in the method is increased. All measures have strengths and weaknesses." The authors would certainly agree that if there were a uniformly best procedure with respect to all criteria including cost and feasibility of widespread implementation, it should be used. Unfortunately there is no such procedure. Choosing a criterion or subset of criteria then implicitly amounts to choosing a winning procedure. Our inquiry is a search for principles for making this choice, given its critical importance. Needless to say we were greatly surprised to find none of the "obvious" criteria emerged as the normative choice. In the spirit of a sensitivity analysis, we would want to know how well a procedure, best by our derived criterion, performed with respect to other criteria. A regression model might be selected using our criterion and subsequently modified to decrease its sensitivity to this choice if that were deemed necessary. But among the "obvious" criteria, average squared error and so on, there is no distinguished element that can serve as such a baseline. There is no basis for choosing to fit a regression model by least squares, for example, rather than average absolute error. So while we agree with the spirit of the referee's remark, we think much more is required before a prescription for inherently subjective choices can be made.

Regression models used in current practice are fitted by least squares even though their performance is measured by other criteria, such as average relative absolute error. The justification for this apparent inconsistency is unknown. In any case, we would propose use of the criterion of Section 2 in both fitting and assessment.

The requirements imposed in Section 3 are the most reasonable of the various alternatives. The condition of *regional independence*, $T_i(S_2, S_1, P_1) = T_i(S_{2i}, S_{1i}, P_{1i})$, $i = 1, \ldots, n$, for example, was ruled out because it forces the model to ignore a vital piece of information, the (assumed) known regional total P_2. If it were admitted, then PI would imply that $T_i = T$ for some T and all i. On top of these conditions *spatial coherence*, $\Sigma \hat{P}_{2i} = T (\Sigma S_{2i}, \Sigma S_{1i}, \Sigma P_{1i})$ for all S_2, S_1, and P_1, would be equivalent to $P_{2i} = \alpha_1 S_{1i} + \alpha_2 S_{2i} + \gamma P_{1i}$ for all i. The result is not inconsistent with RI and TI; however, adding this pair of conditions would lead to $\hat{P}_{2i} = \gamma P_{1i}$, a model which takes no account of the symptomatic variables. Replacing the pair in question successively by one of TC and then TR leads, respectively, to $\hat{P}_{2i} = \alpha(S_{2i} - S_{1i}) + P_{1i}$ and $\alpha(S_{1i} - \beta S_{2i}) + \beta P_{2i}$, $\beta^2 = 1$, according to whether TC or TR is introduced. The efficacy of such linear models is unknown.

ACKNOWLEDGEMENTS

Interest in the problem considered in this paper was stimulated by a problem posed to the second author by the City of Vancouver and by discussions with Mr. Phil Mondor of its Planning Office and Professor H. Hightower, University of British Columbia.

The City of Vancouver supported related research, carried out by the second author and reported elsewhere.

The University of British Columbia and Imperial College, where the work reported herein commenced and finished, respectively, both very generously provided facilities during the period of the leaves of absence of the first and second authors, respectively.

Support for the work was provided by the Natural Sciences and Engineering and the Social Sciences and Humanities Research Councils of Canada, the latter through a Leave Fellowship to the second author.

REFERENCES

Belsley, D. A., Kuh, E., and Welsch, R. E. (1980), *Regression Diagnostics*, Wiley, New York.

Erickson, E. P. (1973), "A Method for Combining Sample Survey Data and Symptomatic Indicators to Obtain Population Estimates for Local Areas," *Demography*, **10**, 137–160.

Erickson, E. P. (1974), "A Regression Method for Estimating Population Changes of Local Areas," *Journal of the American Statistical Association*, **69**, 867–875.

Hauser, P. M. (1981), "The Census of 1980," *Scientific American*, **245**(5), 53–61.

Kitagawa, E. M., et al. (1980), *Estimating Population and Income of Small Areas*, National Academy Press, Washington.

Morrison, P. A., and Relles, D. A. (1976), "A Method of Monitoring Small-Area Population Changes in Cities," *Public Data Use*, **3**, 10–15.

Namboodiri, N. K. (1972), "On the Ratio-Correlation and Related Methods of Subnational Population Estimation," *Demography*, **9**, 443–453.

Nash, J. F., Jr. (1950), "The Bargaining Problem," *Econometrica*, **18**, 155–162.

O'Hare, N. (1976), "Report on a Multiple Regression Method for Making Population Estimates," *Demography*, **13**, 369–379.

Purcell, N. J., and Kish, L. (1979), "Estimation for Small Domains," *Biometrics*, **35**, 365–384.

Rosenberg, H. (1968), "Improving Current Population Estimates through Stratification," *Land Economics*, **44**, 331–338.

Schmitt, R. C., and Crosetti, A. H. (1954), "Accuracy of the Ratio Correlation Method for Estimating Postcensal Population," *Land Economics*, **30**, 279–281.

Swanson, D. A., and Tedrow, L. M. (1984), "Improving the Measurement of Temporal Change in Regression Models for County Population Estimates," *Demography*, **21**, 273–381.

Zidek, J. V. (1982), *A Review of Methods for Estimating the Population of Local Areas*, Technical Report 82-4, Institute of Applied Mathematics and Statistics, University of British Columbia.

PART 3

Theoretical Developments

Using a Covariate for Small Area Estimation: A Common Sense Bayesian Approach

A. P. Dempster and T. E. Raghunathan
Harvard University

ABSTRACT

A case study is presented based on partially hypothetical data where the task is to estimate total wages paid by a type of small business for each of 16 areas. The population size is $N = 436$. A 25% sample is available on Y = wages, while 100% data are available on X = gross business income. Data analysis suggests looking at the transformation $Z = (c + Y)/X$. Several Bayesian analyses using one-way Gaussian random effects models are presented. The effects of outliers, or long tails, in both Z and X are assayed. Comparisons with traditional estimates are displayed.

1. INTRODUCTION

The goals of the present paper are (1) to illustrate an application of Bayesian thinking to small area estimation including covariance adjustment, (2) to contrast our methodology with the more standard approach of basing a choice of method on a comparison of sampling properties of various proposed estimators, and (3) to comment on the type of research effort that would be required to support implementation of our proposed methodology. Our study is limited to a small corner of statistical practice, even a small corner of small area estimation, but we hope that we will encourage further exploration and development of Bayesian methods, which we believe offer conceptually simple and potentially effective alternatives to the cumbersome and inconclusive strategy of relying on either design-based or model-based sampling theory.

Our interest was sparked by reading Hidiroglou et al. (1984) who draw Monte Carlo samples of 500 to compare the sampling properties of a set of standard and nonstandard estimators, assuming 25% samples from four populations each distributed over 18 areas. The raw data for our study is similar to that described in Hidiroglou et al. (1984). The data contain information about 109 establishments in 16 areas on two variables, Y is salaries and wages and X is gross business income.* Our study is not strictly intended to parallel Hidiroglou et al. (1984), because they studied 25% samples from the population of 114 units, while we regard our 109 units as being a 25% sample from a hypothetical population of 436 units. We created X values for the 327 hypothetical units by fitting a truncated Pareto distribution to the observed 109 X values and drawing 327 random X values from the fitted Pareto, which we then distributed at random over the 16 areas so that the 3:1 ratio of unsampled:sampled units was preserved in each area.

In our study we mimic the real-world problem by not knowing the 327 true Y values for the unsampled units. What we mean by *common-sense Bayes* is that we define the task of statistical analysis of the observed units to be the assessment of a posterior distribution over the 327 unknown Y values. For purposes of point estimation we need to be able to assess a posterior expectation for each unknown Y given its X and area identifier, since we may then estimate the total Y for any small area by summing the Y values for those units that are observed and adding the sum of the posterior expectations for those Y values that are not observed, thus obtaining the posterior expectation of the desired target quantity. We may also wish to compute a posterior standard deviation for each estimated area total.

Such point estimates (i.e., posterior expectations) and their posterior standard deviations must be computed from a Bayesian model, which is usually formulated as having two parts, namely, a sampling model given parameters and a prior distribution over parameters. Our position is that both parts of the Bayesian model must be constructed deliberately using available evidence, and both parts represent *prior* knowledge in the technical sense of characterizing uncertainty appropriate *before* the sample data are observed.

---

* We worked entirely from the scatterplot, so lost 5 of the original population of 114 small businesses studied by Hidiroglou et al. (1984) due to a few coincident points. Also, apparently, only 16 of the 18 areas had businesses of this type.

It is traditional in statistics to assess the credibility of a hypothesized sampling model by examining the observed sample data, or even to use the data to suggest an appropriate mathematical form for the sampling model. Since the data are useless for assessing the prior distribution and in fact are only of limited value for resolving doubts about the adequacy of a sampling model, it is necessary to supplement data analysis with sensitivity analysis, not only sensitivity to variations in the prior distribution, but also sensitivity to variations in the sampling model, which the data are unable to detect. When sample sizes are reasonably large, as they often are in the context of government statistics, sensitivity to the prior distribution is often of relatively less concern than sensitivity to failures in the sampling model. Concern for the latter type of sensitivity is shared by common-sense Bayesians and by the robustness school of frequentist statistics, with the difference that our concern is for the sensitivity of posterior means and standard deviations, whereas the frequentist statisticians are concerned with sensitivity of sampling properties of estimators.

We follow the prescriptions implicit in the preceding paragraph by outlining data analysis and model fitting procedures in Section 2, then proceeding to exhibit some limited sensitivity analyses in Section 3. In Section 4, we compare our Bayesian estimates with the results of several traditional estimation procedures, and argue that the results are far from reassuring for proponents of the traditional procedures. Finally, we discuss in Section 5 the implications for research and development of a decision by a statistical agency to make a serious effort to develop common-sense Bayesian methods.

2. DATA ANALYSIS

The plot of salaries and wages (SW) versus gross business income (GBI) is less than ideal because the sample points tend to cluster in the lower left part of the picture. After plotting several versions with transformed Y and X, we decided that a form of display that makes the data easy to see and that suggests a simple class of models is to plot $Z = (c + Y)/X$ against X for various values of c. Figure 1 shows such a plot with $c = 2$. Retaining X as the abscissa serves to remind the viewer of the highly skewed nature of the sample values, while the use of $(c + Y)/X$ serves to remove most apparent correlation between dependent and independent variables. The particular choice $c = 2$ appears to remove all curvature from the relation as well as suggesting a horizontal scatter. The choice of Z in Figure 1 also serves to highlight two outlying values of Z in the

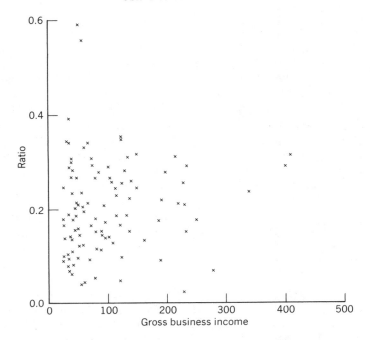

Figure 1. Plot of the ratio $(2 + SW)/\text{GBI}$ versus gross business income.

upper left part of the plot. Finally, note that Z is substantively interpretable as approximately the fraction of gross business income expended as wages and salaries.

We plotted alternative versions of Figure 1, not shown here, which highlighted the 16 areas separately. These plots indicated that systematic differences among the areas are difficult to spot visually against the relatively much more dominant within-area variation, thus suggesting a need for modeling techniques that handle both between- and within-area variation and are appropriate to the situation where between-area effects are relatively small.

Denoting by (Z_{ai}, X_{ai}) the ith sample point in area a, for $i = 1, 2, \ldots, n_a$ and $a = 1, 2, \ldots, 16$, we considered models of the form

$$Z_{ai} = \mu + \beta X_{ai} + \mu_a + e_{ai} \tag{1}$$

where $\mu_1, \mu_2, \ldots, \mu_a$ are regarded as randomly drawn from a normal distribution with mean zero and variance σ_a^2, while the e_{ai} are randomly normal with mean zero and variance σ^2. We also considered variants of the form

$$Z_{ai} = \mu + \beta X_{ai} + \mu_a + \beta_a X_{ai} + e_{ai} \qquad (2)$$

where (μ_a, β_a) are bivariate normal random effects with zero means, but there is little reason either *a priori* or from the data to expect meaningful variance in the slopes $\beta + \beta_a$, so the analyses reported here used the simpler form (1).

The model (1) may be classified technically as a mixed model with fixed effects (μ, β) and random effects (μ_a, e_{ai}). From a Bayesian perspective, (μ, β) may be assigned a diffuse prior distribution because the information in the data is substantial and absent special sources of prior knowledge may be expected to dominate any realistic priors that might be constructed. The more important concern is for the second-order parameters (σ_a^2, σ^2), which control the random effects (μ_a, e_{ai}). Normal Bayesian practice is therefore to integrate out (μ, β) from the likelihood function and look at the restricted likelihood function of (σ_a^2, σ^2). We simplify further by maximizing the restricted likelihood for each contemplated value of the ratio σ_a^2/σ^2. A purer Bayesian method would integrate out rather than maximize, but there is little difference given 109 data points.

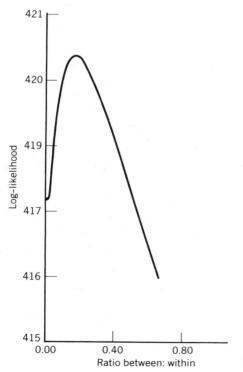

Figure 2. Log-likelihood.

Figure 2 shows a plot of log-likelihood against σ_a^2/σ^2 for model (1) applied to $n = 107$ points, which omit the two outliers visible in Figure 1. We also studied a modification of (1) in which the variance of e_{ai} is inversely proportional to X, specifically, redefining σ^2 so that

$$E(e_{ai}^2) = \frac{\sigma^2}{X} \tag{3}$$

The analysis was repeated for the two choices of $n = 107$ and $n = 109$. The maxima in the four versions of the log-likelihood occur at $\sigma_a^2/\sigma^2 = 0.1538, 0.0041, 0.1481, 0.0033$. The likelihood drops off sufficiently rapidly to rule out $\sigma_0^2 = 0$ or $\sigma_a^2 = \infty$ in each of the four cases. Table 1 shows the estimates of the various fixed and random effects associated with the four fitted models, as well as the estimates of σ_a^2 and σ^2.

As expected, the estimated slope $\hat{\beta}$ is small in the sense of being a relatively small component of a typical estimated $\hat{Z} = \hat{\mu} + \hat{\mu}_a + \hat{\beta}X$.

Table 1.   Estimates

Model	Fixed Effects		Variance Components	
	$\hat{\mu}$	$\hat{\beta}$	$\hat{\sigma}^2$	$\hat{\sigma}_a^2$
$E(\epsilon^2) \propto 1, n = 107$ $\sigma_a^2/\sigma^2 = 0.1538$	0.1840	0.000174	0.006142	0.000945
$E(\epsilon^2) \propto 1, n = 109$ $\sigma_a^2/\sigma^2 = 0.1481$	0.1963	0.000110	0.008528	0.001270
$E(\epsilon^2) \propto \frac{1}{X}, n = 107$ $\sigma_a^2/\sigma^2 = 0.0041$	0.1812	0.000182	0.509621	0.002089
$E(\epsilon^2) \propto \frac{1}{X}, n = 109$ $\sigma_a^2/\sigma^2 = 0.0033$	0.1907	0.000139	0.648099	0.002158

				Random Effects			
$\hat{\mu}_1$	$\hat{\mu}_2$	$\hat{\mu}_3$	$\hat{\mu}_4$	$\hat{\mu}_5$	$\hat{\mu}_6$	$\hat{\mu}_7$	$\hat{\mu}_8$
−0.0069	−0.0199	−0.0045	0.0019	0.0008	−0.0097	0.0308	0.0426
−0.0071	−0.0221	−0.0078	0.0007	−0.0018	−0.0132	0.0564	0.0385
−0.0182	−0.0299	−0.0026	0.0027	−0.0161	0.0053	0.0488	0.0491
−0.0170	−0.0301	−0.0051	0.0012	−0.0168	0.0014	0.0618	0.0430

$\hat{\mu}_9$	$\hat{\mu}_{10}$	$\hat{\mu}_{11}$	$\hat{\mu}_{12}$	$\hat{\mu}_{13}$	$\hat{\mu}_{14}$	$\hat{\mu}_{15}$	$\hat{\mu}_{16}$
−0.0181	−0.0129	−0.0286	0.0261	0.0179	−0.0202	0.0143	−0.0136
−0.0098	−0.0157	−0.0323	0.0219	0.0167	−0.0205	0.0099	−0.0139
−0.0284	−0.0156	−0.0359	0.0429	0.0463	0.0052	0.0275	−0.0811
−0.0219	−0.0176	−0.0382	0.0374	0.0456	0.0064	0.0235	−0.0737

Area-to-area effects represented by $\hat{\mu}_a$ are rarely more than 2% of the typical $\hat{\mu}$. The $\hat{\mu}_a$ are similar for Models 1 and 2 and for Models 3 and 4, but differ between pairs of cases. The overall $\hat{\mu}$, on the other hand, are similar for Models 1 and 3 but about 5% higher for Models 2 and 4, showing the effect of including outliers.

3. SENSITIVITY ANALYSES

Each of Models 1 to 4 implies a posterior expected total Y value for each area and an associated posterior standard deviation. These are summarized in Table 2. We also computed estimated area totals for four parallel models with $\sigma_a^2/\sigma^2 = 0$ and four parallel models with $\sigma_a^2/\sigma^2 = \infty$. The first set of four corresponds to setting area effects equal to zero, whereas the second set of four treats the area parameters μ_a as fixed effects to be estimated strictly from each area with no shrinking to means across areas. These extreme cases are included as benchmarks even though the Figure 2 displays render them highly implausible. A more ambitious form of Bayesian analysis that would average across a posterior for σ_a^2/σ^2 was not attempted due to limited resources of time.

We illustrate two ways to assess sensitivity to model change from the numbers in Table 2. The first method summarizes the percentage changes in area total estimates corresponding to model changes. Choosing Model 1 as a standard, we computed the average percentage change and the root mean square percentage change from Model 1 to 11 other models, namely, Models 2, 3, 4, and the modifications of all four models to $\sigma_a^2/\sigma^2 = 0$ and ∞. The systematic changes are small, typically a fraction of 1%, but the typical individual change among Models 1–4 is about 5%. We judge these values large enough to provoke concern about choosing a good consensus model, especially if allocations of funds to areas is at stake.

The second way to assess sensitivity is relative to posterior standard deviation. Again we take Model 1 as a standard, and we compute the average change and root mean square change from Model 1 to each of the same 11 models as before, where change is measured in units of Model 1 standard deviation and average is taken over the 16 areas. The changes among Models 1–4 are typically only a fraction of one standard deviation, thus suggesting that there is little information in the sample data that would point conclusively to a choice among these models.

Since posterior expectations are available for each population unit, we can easily estimate population totals for subsets of the population. For example, by computing such estimates for the subset of each area with

Table 2. Area Total Estimates and Their Posterior Standard Deviations

Area	Model 1	Model 2	Model 3	Model 4	Method 1	Method 2	Method 3
1	152.2923 (8.6158)	158.1292 (9.9969)	150.6502 (11.1135)	155.4586 (11.8039)	77.8200	94.9483	66.6284
2	326.7614 (15.4496)	332.9072 (17.9828)	324.3244 (19.0801)	330.1212 (20.0858)	362.0000	374.4055	263.1380
3	313.5624 (46.5646)	311.3648 (54.1151)	295.8064 (54.9607)	298.0519 (58.2297)	110.0000	316.1479	278.5408
4	71.5210 (9.0045)	70.0092 (10.5293)	67.7400 (8.7301)	67.6255 (9.5794)	29.6920	75.3070	69.8184
5	722.4649 (84.7277)	704.5467 (98.8945)	725.5067 (103.4254)	716.4043 (110.3925)	547.0000	698.8704	640.0263
6	454.8848 (61.1114)	454.5677 (70.9501)	453.6441 (85.0145)	454.7894 (88.2278)	278.0000	452.1350	422.9736
7	502.5367 (36.6678)	504.3511 (42.8489)	471.6581 (37.3919)	476.7943 (41.1851)	779.0000	451.9878	696.0690
8	624.1407 (69.4551)	604.2452 (81.2187)	657.9801 (73.1358)	644.2694 (79.4690)	562.0000	561.0560	692.1822
9	2907.1177 (287.1483)	3121.7645 (329.0271)	3081.8928 (261.2997)	3193.8314 (281.8732)	1691.0000	2421.0166	2191.9688
10	380.3147 (33.1689)	378.3877 (38.7435)	387.3073 (35.9391)	382.7899 (39.3211)	220.0000	343.6261	269.9536

11	971.7279 (80.5664)	1002.7566 (94.0261)	921.6235 (63.2379)	952.1459 (70.2028)	588.0000	1061.3354	814.3331
12	699.4058 (77.2591)	688.9406 (90.2285)	688.6146 (86.9633)	684.8312 (94.2995)	566.0000	673.6256	785.6279
13	412.2247 (30.3190)	403.3022 (35.5292)	402.0206 (27.5881)	398.0316 (30.5217)	746.0000	369.4890	611.3897
14	695.9605 (49.7158)	690.8639 (58.1166)	728.8479 (53.8442)	722.0926 (58.9768)	886.0000	602.0639	725.3632
15	1117.4982 (105.1819)	1117.4974 (122.6903)	1227.8171 (101.4617)	1233.1227 (111.0264)	954.0000	995.8535	1120.4792
16	93.5468 (11.6682)	95.9658 (13.6151)	105.4474 (11.5324)	108.4614 (12.6968)	51.6400	149.3402	-6.3480
Average scaled percentage change from model 1	0.0000	0.0060	0.0868	0.0328	-1.2240	-0.4627	-1.1420
Root mean square scaled percentage change from model 1	0.0000	0.4002	0.4276	0.4995	5.0567	2.3997	4.2249

$250 \times 10^3 < X < 500 \times 10^3$ we can perform sensitivity analyses that directly look at the long upper tail of the distribution of X where sensitivity should be increased. Sensitivity analysis, analogous to that described in the preceding two paragraphs, was made using 40 units in our population of 436 with $X > 250 \times 10^3$. The systematic changes were found to be generally larger than those corresponding to the whole population, but the increases in the sensitivity are not dramatic.

The sensitivity analyses reported here are certainly not exhaustive. Given adequate time and facilities we would have conducted analyses for models with slope parameter varying from area to area, and also for regression models with quadratic curvature. Also, we regard our approach to outliers (i.e., repeating the analysis with and without them) to be inadequate. A preferable method would be to fit models with skewed distributions for the e_{ai}, but the effort required is beyond the scope of the present study.

4. COMPARISONS WITH OTHER ESTIMATORS

We report here comparisons of the Model 1 Bayes estimator, which we used above as a standard, with three of the estimators studied in Hidiroglou et al. (1984). Method 1 is the "expansion" or "poststratified" estimators of those authors. These estimators are the same in our situation because the sampling fraction is 25% in all areas. In effect, Method 1 ignores the covariate X, and simply uses the sample mean $\bar{Y}_a$ in each area multiplied by the population size N_a to estimate population total. Method 2 is the "ratio-synthetic" estimator, which computes population totals X_a for each of the 16 areas, and scales these by the same factor

$$b = \left(\sum_{a=1}^{16} \sum_{i=1}^{n_a} Y_{ai} \right) \Big/ \left(\sum_{a=1}^{16} \sum_{i=1}^{n_a} X_{ai} \right) \tag{4}$$

Method 3 is the "regression-ratio" estimator proposed by Särndal (1981), which starts from the Method 2 estimate and adds the area specific adjustment

$$\frac{N_a}{n_a} \left(\sum_{i=1}^{n_a} \{ Y_{ai} - b X_{ai} \} \right) \tag{5}$$

where b is obtained from (4). Hidiroglou et al. (1984) also studied several "empirical Bayes" estimators based on the methodology of Fay and

Herriot (1979). We omitted these from our study because they rely on the between-area regression of Y_a on X_a, whereas it is evident from inspection of the data that the major gains from covariance adjustment are related to within-area correlation.

Table 2 displays estimated area totals for Y using Methods 1, 2, and 3 along with the four Bayesian models explored in this paper. Average and root mean square percentage changes over 16 areas from Model 1 for Methods 1, 2, and 3 were computed. Percentage changes were dramatically higher. Method 2 fared better with an average percentage change 2.5% with root mean square 19.9%.

5. DISCUSSION

The Bayesian paradigm is not just one more set of principles that can be used to derive estimators. If a consensus can be constructed around a Bayesian model, or around a class of models, such that conclusions are insensitive to varying the model within the class, then the Bayesian conclusions are *the correct conclusions*. For example, if squared error loss is roughly appropriate, then the use of the posterior expectation as the estimator is virtually dictated. We believe that in many situations a consensus sufficient unto its time and place can be constructed, and, if so, the use of other estimators is risky.

The use of Methods 1 and 2 of Section 4 is questionable not because they are non-Bayesian, but because their Bayesian interpretations relate to unrealistic models. To use Method 1, a Bayesian would need to believe that knowledge of X is of no value in predicting Y. Such a belief is clearly untenable given a quick inspection of Figure 1. Method 2 can be given a Bayesian interpretation if one believes that the regression of Y on X is represented by a line through the orion, that the deviations of Y from the line are roughly normally distributed with a variance proportional to X, and that there is no systematic area-to-area variation. The first of these three assumptions is roughly borne out by Figure 1, but the second is clearly false so that (4) is an inefficient estimate of the slope of the line, and the data analysis of Section 2 strongly suggests that some area-to-area variation is present. Despite the weaknesses of the second and third assumptions, Method 2 is clearly superior to Method 1 for this particular data set, but surely much less plausible than any of the four Bayesian estimates studied in Sections 2 and 3. Method 3 is less easily fitted into a Bayesian model, because it shares with Method 2 a reliance on (4) for regressing Y on X and then attempts an ad hoc adjustment for area effects whose Bayesian credentials are not transparent. We are therefore not surprised that Method 3 appears to be inferior to Method 2, even though modification of Method 2 to allow systematic area effects is desirable. Note that our models incorporate differences among area means on $Z = (2 + Y)/X$, so in effect incorporate different Y/X slopes whose necessity is backed up by the plot in Figure 2.

Our very limited case study points to two types of expertise that need to be developed before Bayesian small area estimation can move into the realm of practice. First, as is evident from the case study, there is a need for data analysis keyed to the special needs of model construction for Bayesian inference. Second, there is a need to develop a tool kit of models that are likely to be usable across a range of practical situations. This tool kit includes mathematical descriptions, but also, and more critically, the tool kit must provide algorithms and software for implementing Bayesian methods, including the associated requisite sensitivity analyses.

Statistical research has traditionally been organized around methods and the performance of methods under sampling models. While such studies can provide many important insights, our analysis suggests that an increasing portion of research resources should be shifted to case studies that will provide a framework for the scientifically demanding task of developing situation-specific models at a level of both relevance and detail sufficient to support credible Bayesian inferences. Note that both models for random samples and models for nonsampling errors are required in typical cases. Any realistic extension of our case study would quickly become very complex if attempted on a national scale, including more classes of employment and more covariates defined over sampling frames obtained from a range of census and administration sources. The very complexity will provide, however, opportunities for assessing sources of variability whose incorporation into adequate models is highly desirable. The data analytic and related modeling tasks will be very challenging, but there is no other way to proceed if estimation is to progress beyond the narrow rule-based outlook that currently holds sway.

Finally, a few words on computing are in order. There are obviously major logistical tasks involved in assembling, controlling quality, and providing convenient access to large data bases. Beyond this, because distributed computing is becoming so powerful and inexpensive, old limits on the feasibility of complex estimation techniques are no longer meaningful. The guiding principle with technology is that it should be appropriate for the task at hand, neither technology in search of a problem for its own sake nor outmoded technology serving only to limit progress. We believe that there is a good match between the statistical need for more focused estimation based on Bayesian principles and the computing technologies required to support production of such estimates. A major need is for research talent able to conceive and develop the required software.

APPENDIX: COMPUTATIONAL DETAILS

Let Z_a be a $n_a \times 1$ vector of transformed values $(2 + Y)/X$ for area a. Let α be a $p \times 1$ vector of fixed effects, β_a be a $q \times 1$ vector of random effects for area a. Let T_a and S_a be a $n_a \times p$ and $n_a \times q$ matrix of covariates linking fixed effects and random effects to Z_a. For each area, we write our model as

$$Z_a = T_a \alpha + S_a \beta_a + \epsilon_a$$

where $\epsilon_a \sim N(0, \sigma^2 I)$ and $\beta_a \sim N(0, D)$, D a positive-definite matrix of order q. Let

$$Z^t = (Z_1^t, Z_2^t, \ldots, Z_{16}^t)^t$$

$$T^t = (T_1^t, T_2^t, \ldots, T_{16}^t)^t$$

$$S = \begin{bmatrix} S_1 & 0 & \ldots & 0 \\ 0 & S_2 & \ldots & 0 \\ \vdots & \vdots & & \vdots \\ 0 & 0 & \ldots & S_{16} \end{bmatrix}$$

where the superscript t stands for matrix transpose.

In order to estimate the fixed effects α and the random effects β_a where $a = 1, 2, \ldots, 16$, we first form the matrix of sum of squares and products

$$\begin{bmatrix} T'T & T'S & T'Z \\ & S'S + \sigma^2 D^{-1} \otimes I & S'Z \\ & & Z'Z \end{bmatrix} \tag{6}$$

where $\otimes$ denotes Kronecker product, and second perform sweep operations on the block of indices $1, 2, \ldots, p + 16q$. The sweep operations are described in Dempster (1969, 1982) and Goodnight (1979). Since sweep operations commute and we need to perform sweep for various choices of $K = \sigma^2 D^{-1}$, our algorithm sweeps in two phases. First over the block of indices $1, 2, \ldots, p$ and second over the block indices $p + 1, \ldots, p + 16q$

for various choices of σ^2 and D. An important quantity in calculating the log-likelihood is the determinant of the matrix $Q = S^t R S + \sigma^2 D^{-1} \otimes I$ where $R = I - T(T^t T)^{-1} T$. This is simply obtained by multiplying pivotal quantities, during each stage of the second phase of sweeping operations.

In the last column of the resulting matrix, first p rows contain the estimates of the fixed effects, row $p + 1$ through $2p + 16q$ contain estimates of random effects. The estimate $\hat{\sigma}^2$ of σ^2 for the fixed ratio $\sigma^2 D^{-1}$ is obtained by dividing the last element of the last column by $n - p$, where n is the total sample size. The log-likelihood is then calculated using the expression

$$\log L = - \frac{n - p}{2} \log \hat{\sigma}^2 - \tfrac{1}{2} \, \text{logdet} \ Q + 8 \ \text{logdet} \ K$$

Our algorithm will work well if the matrix (6) is not close to singular. Further note that we need to store only the upper half of the matrix (6).

ACKNOWLEDGMENT

This work was facilitated in part by National Science Foundation Grant DMS-8201820.

REFERENCES

Dempster, A. P. (1969), *Elements of Continuous Multivariate Analysis*, Addison-Wesley, Reading, MA.

Dempster, A. P. (1982), "Some formulas useful for covariance estimation with Gaussian linear component models," in *Statistics and Probability: Essays in Honor of C. R. Rao*, (G. Kallianpur, P. R. Krishnaiah and J. K. Ghosh, eds.), Amsterdam, North-Holland, pp. 213–220.

Fay, III, R. E., and Herriot, R. A. (1979), "Estimates of income for small places: An application of James Stein procedures to census data," *Journal of the American Statistical Association*, **74**, 405–410.

Goodnight, J. H. (1979), "A tutorial on the sweep operator," *American Statistician*, **33**, 149–158.

Hidiroglou, M. A., Morry, M., Dagum, E. G., Rao, J. N. K., and Särndal, C. E. (1984), "Evaluation of small area estimators using administrative records," *Proceedings of the American Statistical Association, Social Statistics Section*.

Särndal, C. E. (1981), "Frameworks for inference in survey sampling with applications to small area estimation and adjustment for non response," *Bulletin Int. Stat. Inst.*, **49**(1), 494–513 (*Proceedings, 43rd Session, Buenos Aires*).

Application of Multivariate Regression to Small Domain Estimation

R. E. Fay
U.S. Bureau of the Census

ABSTRACT

The original development and most applications and extensions of the linear regression approach to estimation for small domains have been restricted to a univariate perspective. In this formulation, an estimated univariate Y for the target variable available for a number of small domains from a sample survey is related to one or more predictor variables X obtained from independent sources for these same domains. In some applications, however, auxiliary information Z on the variable Y may arise from the same sample survey. The treatment of Z as part of the independent variables X in standard linear regression may give misleading estimates for the domains, depending on the nature of the sampling covariances between Y and Z. Instead, viewing the problem as multivariate linear regression for the combined dependent vector (Y, Z) may lead to a more correct formulation of the problem, depending on the application. Recent research on components of variance models for multiple regression suggests approximate empirical Bayes versions. Under some circumstances, the resulting small domain estimates will be far more powerful than if Z were ignored. These notions are illustrated by an application to data from the U.S. Census Bureau.

1. INTRODUCTION

Linear regression and its variants are well-known approaches to small domain estimation. Although scattered precedents for its use appeared earlier, the work of Erickson (1973, 1974) was instrumental in developing the theory and illustrating the applications of such methods to small domain estimation. The paper of Purcell and Kish (1979) describes applications of this approach and compares its features with alternatives. Numerous other efforts have followed, although no enumeration of them will be attempted here.

In the typical situation, estimates, Y, are available from a sample survey of some characteristic of interest for a number of domains. Often, the sample estimates Y are regarded as unbiased for practical purposes but are limited by large sampling variability for some or all of the domains. Occasionally, some individual domains of interest may lack a corresponding sample estimate altogether. Generally, the regression approach to small domain estimation employs independent variables X to develop a regression model fitted at the domain level and substitutes the fitted values for the domain estimates. Embellishments on this approach, such as those in Fay and Herriot (1979), who adapted results of James and Stein (1961) and of Efron and Morris (1972, 1973, 1975), may reincorporate the sample estimates into the estimation for the domains or otherwise constrain the estimates under the models.

Formulation of such models has included error terms for both the sampling variation present in Y and lack of fit of the model to the underlying true or "census" values. The values of the independent variables are generally not treated as stochastic, however. For example, this approach would seem appropriate when the independent variables are obtained from administrative sources. Some or all of the components of X may come from a much larger sample survey, but these values may be treated as nonstochastic when the effect of sampling error on X is negligible relative to that on Y. The primary example of this sort is the inclusion of census data among the independent variables to predict a variable Y obtained from an independent sample survey, since most census characteristics in the United States and other countries are based on very large samples rather than complete enumerations.

The consequences of the presence of nontrivial sampling variances in components of the X's are varied. If the predictors subject to sampling variance are denoted by Z and the nonstochastic predictors are denoted by X, then the usual regression approach may prove statisfactory if:

(i) Z is independent of Y or uncorrelated with it; and

(ii) the effect of sampling variance in Z on the prediction can be conveniently subsumed into the assumed structure of the lack of fit of the regression model to the underlying unknown "census" values, y, for the complete population.

If sampled values Z of the predictor variables are used to fit the regression equation, but if predicted values for the model are instead to be computed from corresponding population values z, presumed known, then more formal treatment of the effect of sampling variance in Z is required. This circumstance is thus an instance of the errors-in-variables problem, and estimation of the regression model using an errors-in-variables formulation would appear more suitable for developing estimates based on population values z.

In other circumstances, however, only the sampled values of Z may be available for prediction. If:

(i) Y and Z have correlated sampling errors or
(ii) the sampling errors of (Y, Z) are not compatible with the presumed structure of the model errors in the underlying population,

then a different formulation is necessary. To illustrate the second of these circumstances, the underlying model for the superpopulation variance of the population quantities (y, z) given X may assume a constant variance–covariance structure over the domains of interest, while the sampling errors of (Y, Z) may vary considerably over the same domains. Use of standard linear regression of Y on (X, Z) will misrepresent the special nature of this error structure.

This paper will illustrate the formulation of the estimation problem as a multiple linear regression of (Y, Z) on X, using a components of variance model of the form reviewed by Harville (1977), and developed by many authors, including Hartley and Rao (1967), Harville (1976), and Searle (1971). Fuller and Harter (1985) provide a general discussion of the application of components of variance models to small domain estimation.

Section 2 will summarize an estimation problem for which this approach is suggested and includes a summary of a simple linear regression model that was the predecessor. Section 3 attempts to motivate the multivariate approach to estimation by showing in an elementary example that, even when only one component of a vector is of interest, estimating the joint vector may improve the estimation of a given component compared to restricting attention only to the single component. On this basis, Section 4 characterizes the presumed model for this application. Section 5 describes the remaining research and adds concluding remarks.

2. THE ESTIMATION OF MEDIAN INCOME FOR FOUR-PERSON FAMILIES BY STATE

The U.S. Department of Health and Human Services (HHS) administers a program of energy assistance to low-income families. Eligibility for the program is determined by a formula whose most important variable is an estimate of the current median income for four-person families by state (the 50 states and District of Columbia).

The annual demographic supplement to the March sample of the Current Population Survey (CPS), the monthly labor force survey in the United States, provides national statistics on families and income, including the median income of four-person families for the preceding year. Separate estimates of median income for four-person families are also available by state, but to considerably less reliability than the national estimate. The coefficient of variation (cv, standard error divided by estimate) of this statistic is as small as 2 or 3% in a few states but is in the range of 6–10% for many states. This level of variation appeared too high to administrators of the program, who wished a more smoothly behaving set of estimates from year to year. Compared to many problems of small domain estimation, however, the sample estimates are relatively reliable. By comparison, the CPS is specifically designed to produce estimates of annual average unemployment with a cv of 10% in small states and with somewhat greater reliability in large states, and these sample estimates have been published and used directly.

By an informal agreement, the Census Bureau has produced estimates of the median income for four-person families by state for HHS through a linear regression methodology since the latter part of the 1970s. Several parallels may be noted to the methods discussed by Fay and Herriot (1979). In this method, sample estimates Y of the state medians for the most current year, c, are first obtained, along with estimates of their variances. These estimates become the dependent variable in a linear regression using the single predictor variable

$$\text{Adjusted census median}(c) = [\text{BEA PCI}(c)\,/\,\text{BEA PCI}(b) \\ \times \text{Census median}(b) \qquad (1)$$

where Census median(b) represents the median income of four-person families in the state for base year b from the most recently available decennial census, BEA PCI(c) represents estimates of per capita personal income produced by the Bureau of Economic Analysis of the U.S. Department of Commerce for the current year c, and BEA PCI(b) represents the corresponding estimate for the same base year b as the census. Formula (1) thus attempts to adjust the preceding census median by the proportional growth in BEA PCI.

This regression predictor was selected on the basis of its success in predicting the 1970 census values of 1969 median income on the basis of the 1960 census values for 1959. In developing this model, the states were divided on the basis of population into four groups of 12 or 13 states each. Model variances for each were computed as the average of squared residuals of 1969 census medians fitted by (1) and a constant term. The two groups of larger states were observed to have smaller estimated model errors than the smaller states.

Two final steps were incorporated into the estimation. One was the averaging of the CPS sample of the current median income with the regression estimate, weighting the regression estimate inversely proportional to the average model error obtained from the fit of the census values for 1969 and weighting the sample estimate inversely proportional to its sampling variance. This combining of estimates parallels a similar operation in the estimator described by Fay and Herriot (1979), but in this case is more strictly Bayesian than empirical Bayes since estimates of the model variance were not constructed from the current CPS sample values.

The second step constrained the resulting value of the average of the sample and regression estimates to deviate by no more than one standard deviation, that is, the square root of sampling variance, from the sample value by pulling all estimates beyond the constraint to one standard deviation away. This second step also parallels the estimator described by Fay and Herriot (1979) and is based on results of Efron and Morris (1972). The purpose of this second step was to limit the risk to any one state, no matter how deviant the true value for the state from the fitted regression value.

A review of this model was undertaken once median income of four-person families by state in 1979 became available from the 1980 census. Several conclusions emerged from this review:

(i) The estimated model error from the fitting of 1969 medians from the census from 1959 medians reasonably approximated the level of actual model error for 1979, but the differentials observed earlier between groups by size greatly exaggerated more current differences. Accordingly, it now appeared appropriate to consider a single level of model error across all states.

(ii) The predictive value of the regression appeared usefully improved if, in addition to (1), Census median(b) were also included as a second independent variable. The effect of this inclusion is to adjust for possible overstatement of the effect of change in BEA income upon the median incomes of four-person families, by allowing, in a sense, for a "regression toward the mean" in the actual rate of growth compared to the prediction relying on BEA estimates through (1).

(iii) The constraint of one standard deviation from the sample estimate in fact did not reduce the number of extreme errors compared to the unconstrained weighted average of the sample and regression estimates. This apparent failure is not because the constraint would not function as required in the presence of severe model errors. Rather, it reflects the absence of severe model errors for any states.

Although the performance of the model generally met the expectations based on the previous performance for the 1959–1969 decade, and incorporation of the first two of the three preceding points would presumably result in improvements for current application, staff of the office in HHS using these estimates requested that further improvements be made. Specifically, during the current period of moderating inflation in the United States, the model was now more likely to show a decrease in median income for individual states from one year to the next than during the preceding period of more rapid inflation. One cosmetic disadvantage to the decreases was that they provided visible evidence that the estimates were based on sample data and not perfect. The more serious problem, however, was that, in administering an assistance program under the given formula, when the state's median declined, some persons qualified in the first year might not be qualified in the next year at the same income. In some cases, such decreases might have been indeed real, particularly when the estimates from BEA themselves showed declines. Many decreases, however, resulted from wide swings in the state's sample estimates over the two years, often, but not always, affecting the application of the constraint on the estimate to one standard error in one or both years.

A possible improvement in the estimation might result from the observed fact that median incomes in 1979 and 1969 by state from the censuses for three-, four-, and five-person families shows a strong statistical relationship among the three sets. For example, in 1979, the smallest ratio of the four-person median to the three-person median in any one state is 1.072 and the highest 1.162. Two-thirds of the states fall between 1.089 and 1.135. Furthermore, some of the differences between states appears structural, since almost all of the higher ratios are confined to states with a high proportion of Black population. The relationship between medians for four- and five-person families is nearly as strong.

If exact figures were available for the medians of three-person families by state, they could be used as an ideal independent variable for a regression model for four-person families. Of course, no such figures are available outside the years in which the census is taken. The CPS does produce sample estimates, however. Although this question will be studied further in the course of our research, there is presumably only a slight sampling correlation between the CPS estimated medians for three-, four-, and five-person families.

Section 4 will describe a model, based on a components of variance analysis, to incorporate the estimated three- and five-person medians into the estimation. These two variables are added as additional dependent variables in the model, although the objective is to improve the estimates for four-person families. Since improving a model for a given dependent variable by increasing the number of dependent variables may seem surprising to many, the next section offers an elementary example of how this may occur.

3. AN ELEMENTARY EXAMPLE OF MULTIVARIATE ESTIMATION

To illustrate that multivariate estimation of a vector may improve the estimates of the components, let $y_1 = (u_1, v_1)' \sim N(\mu, D)$, and $y_2 = (u_2, v_2)' \sim N(\mu, A)$ be two independent random variables with common expected values $\mu = (\mu_1, \mu_2)'$. No specific relationship is assumed between μ_1 and μ_2, but D and A will both be considered nonsingular.

In addition to the general case, an illustrative special case will be specifically examined: $D = I$, and A has the form

$$A = \begin{pmatrix} \sigma^2 & \rho\sigma^2 \\ \rho\sigma^2 & \sigma^2 \end{pmatrix}$$

If only u_1 and u_2 were observed, the maximum likelihood (MLE) and best linear unbiased (BLUE) of μ_1 is given in the general case by

$$\mu_1^* = (D_{11}^{-1}u_1 + A_{11}^{-1}u_2)/(D_{11}^{-1} + A_{11}^{-1}) \tag{2}$$

and in the specific case by

$$\mu_1^* = (u_1 + \sigma^{-2}u_2)/(1 + \sigma^{-2}) \tag{3}$$

where A_{11} and D_{11} are the $(1, 1)$th elements of the respective covariance matrices. Based on the full vectors y_1 and y_2, however, the MLE and BLUE estimators of μ are given by

$$\mu^* = (D^{-1} + A^{-1})^{-1}(D^{-1}y_1 + A^{-1}y_2) \tag{4}$$

Note the similarity between the general expressions for the univariate, (2), and multivariate, (4), cases.

The first component $(\boldsymbol{\mu}^*)_1$ of $\boldsymbol{\mu}^*$ in this hypothetical case is

$$(\boldsymbol{\mu}^*)_1 = \frac{[\sigma^2(1-\rho^2)+1]u_1 + \rho v_1 + (1+\sigma^{-2})u_2 - \rho v_2}{\sigma^2(1-\rho^2) + 2 + \sigma^{-2}} \tag{5}$$

Note that in (5) the sum of the coefficients on the two terms with expected value μ_1, namely, u_1 and u_2, is unity, and that the sum of coefficients on the remaining two terms is zero, consistent with unbiasedness.

For $\rho = 0$, (5) reduces to (3) after appropriate algebraic manipulation. Generally, if $\mathbf{A}$ is a scalar multiple of $\mathbf{D}$, $(\boldsymbol{\mu}^*)_1$, the first component of $\boldsymbol{\mu}^*$ from (4), will be equal (2). In other words, when $\mathbf{A}$ and $\mathbf{D}$ are scalar multiples of each other, there is no advantage to considering the multivariate estimation problem in trying to estimate individual components.

As ρ approaches unity in the specific example, the estimate (5) differs considerably from (3). For example, when $\sigma^2 = 1$, (3) weights u_1 and u_2 equally, while the estimator (5) approaches weights of $\frac{1}{3}$ for u_1 and v_1, $\frac{2}{3}$ for u_2, and $-\frac{1}{3}$ for v_2. In other words, (5) approaches an estimate based on the simple average of three quantities: u_1, u_2, and $v_1 - (v_2 - u_2)$. The last of these three quantities, $v_1 - (v_2 - u_2)$, is a third unbiased estimator of μ_1, which exploits the small sampling variance of $v_2 - u_2$ as ρ approaches unity to give an estimator of reliability approaching that of u_1 or u_2 in this special case.

These observations suggest the general means by which multivariate linear regression may improve the estimation of a given component. In the application to the estimation of medians for four-person families, the estimate y_1 becomes analogous to sample estimates of median incomes for four-, three- and five-person families. The three sample estimates are presumably virtually uncorrelated, since they represent statistics for different households in the sample. Thus, the sampling variances of the medians should approximate a diagonal matrix $\mathbf{D}$. Because the actual medians are highly related in the population, however, the residuals from any regression model for the population values should have a high degree of correlation, represented by the matrix $\mathbf{A}$. The next section gives a formal definition of the intended model.

4. THE MODEL FOR FAMILY MEDIANS BY STATE

Let $\boldsymbol{\mu}_{(i)}$ represent the true medians of four-, three-, and five-person households in state i. For each component of $\boldsymbol{\mu}_{(i)}$, assume that $\boldsymbol{\mu}_{(i)j}$ has been sampled from a superpopulation distribution with expected value $\mathbf{X}_{(i)(j)}\boldsymbol{\beta}_j$, for $\mathbf{X}_{(i)(j)}$ a row vector containing the constant 1, the value of the respective median from the preceding census, and the updated median for the family size in the form (1). Assume also that the error terms about the expected values in the superpopulation are jointly distributed as $N(\mathbf{0}, \mathbf{A})$. The errors are assumed independent between states.

The corresponding CPS estimates $\mathbf{Y}_{(i)}$ for state i are assumed $N(\boldsymbol{\mu}_{(i)}, \mathbf{D}_{(i)})$, with known covariances $\mathbf{D}_{(i)}$. Again, the sampling errors will be assumed independent across states.

Under this superpopulation model, the unconditional distribution of the $\mathbf{Y}$'s may be expressed as a mixed components of variance model. Setting $\mathbf{Y} = (\mathbf{Y}_{(1)1}, \ \mathbf{Y}_{(1)2}, \ \mathbf{Y}_{(1)3}, \ \mathbf{Y}_{(2)1}, \ldots, \mathbf{Y}_{(51)3})^T$, and $\mathbf{X}$ to be the matrix whose first rows are given by

$$\mathbf{X} = \begin{pmatrix} \mathbf{X}_{(1)1} & \mathbf{0} & \mathbf{0} \\ \mathbf{0} & \mathbf{X}_{(1)2} & \mathbf{0} \\ \mathbf{0} & \mathbf{0} & \mathbf{X}_{(1)3} \\ \mathbf{X}_{(2)1} & \mathbf{0} & \mathbf{0} \\ & \vdots & \end{pmatrix}$$

the model is

$$\mathbf{Y} = \mathbf{X}\boldsymbol{\beta} + \mathbf{b} + \mathbf{e} \tag{6}$$

where $\mathbf{b}$ represents random effects arising from model errors with presumed covariance matrix in the block diagonal form

$$\mathbf{A}^* = \begin{pmatrix} \mathbf{A} & \mathbf{0} & \mathbf{0} & \mathbf{0} & \cdots \\ \mathbf{0} & \mathbf{A} & \mathbf{0} & \mathbf{0} & \cdots \\ \mathbf{0} & \mathbf{0} & \mathbf{A} & \mathbf{0} & \cdots \\ \mathbf{0} & \mathbf{0} & \mathbf{0} & \mathbf{A} & \cdots \\ & & \vdots & & \end{pmatrix}$$

and $\mathbf{e}$ denotes sampling errors with covariance matrix

$$\mathbf{D} = \begin{pmatrix} \mathbf{D}_{(1)} & \mathbf{0} & \mathbf{0} & \cdots \\ \mathbf{0} & \mathbf{D}_{(2)} & \mathbf{0} & \cdots \\ \mathbf{0} & \mathbf{0} & \mathbf{D}_{(3)} & \cdots \\ & \vdots & & \end{pmatrix}$$

If **A** were known, the BLUE estimate of the fixed effects would be given by

$$\hat{\boldsymbol{\beta}} = [\mathbf{X}^T(\mathbf{D} + \mathbf{A}^*)^{-1}\mathbf{X}]^{-1}\mathbf{X}^T(\mathbf{D} + \mathbf{A}^*)^{-1}\mathbf{Y} \tag{7}$$

The objective is to estimate the true medians

$$\boldsymbol{\mu} = \mathbf{X}\boldsymbol{\beta} + \mathbf{b}$$

that is, a linear combination of fixed and random effects. The BLUE of $\boldsymbol{\mu}$ is given by (Harville, 1976):

$$\hat{\boldsymbol{\mu}} = \mathbf{X}\hat{\boldsymbol{\beta}} + \mathbf{A}^*(\mathbf{D} + \mathbf{A}^*)^{-1}(\mathbf{Y} - \mathbf{X}\hat{\boldsymbol{\beta}}) \tag{8}$$

Equation (8) gives the BLUE as the regression equation plus the multivariate residual $(\mathbf{X} - \mathbf{X}\hat{\boldsymbol{\beta}})$ times a shrinkage factor $\mathbf{A}^*(\mathbf{D} + \mathbf{A}^*)^{-1}$. This estimator corresponds to the posterior mean of the analogous Bayes model with a uniform prior on the fixed effects.

The log-likelihood, ignoring additive constants, is

$$L(\mathbf{A}, \boldsymbol{\beta}; \mathbf{Y}) = -\tfrac{1}{2} \log[\det(\mathbf{A}^* + \mathbf{D})] - \tfrac{1}{2}(\mathbf{Y} - \mathbf{X}\boldsymbol{\beta})^T(\mathbf{D} + \mathbf{A}^*)^{-1}(\mathbf{Y} - \mathbf{X}\boldsymbol{\beta}) \tag{9}$$

By choosing test values of **A**, computing $\hat{\boldsymbol{\beta}}$ according to (7), and then the log-likelihood from (9), one may compute the maximum likelihood estimate for this problem through a search procedure on test values of **A** or by algorithmic methods such as those reviewed in Harville (1977). If the maximum likelihood estimator of **A** is used in (7) and (8), (8) becomes, broadly speaking, a parametric empirical Bayes procedure in the sense of Morris (1983).

5. REMAINING RESEARCH

The model and estimation strategy described in the preceding section have not yet been fully implemented, but preliminary calculations on estimates from the 1984 CPS appear promising. The section will briefly state the research required to complete this effort.

Before the calculations can be completed, more refined estimates of the sampling variances and especially covariances will be required. The estimates are in fact extrapolated rather than exact medians, but the known smoothness of the income distributions at the state level can be employed to advantage in developing an estimator of the sample variances and covariances.

Although general programs such as SUPER CARP (Hidiroglou, Fuller, and Hickman, 1978) may often be of help in estimating mixed models such as that described, a specialized FORTRAN program will be written to implement the computations.

In addition to the improvements arising from the multivariate treatment of this problem, special attention will be directed toward the question of application of the one standard error constraint. The strong model correlation between the medians by family size could be suitably reflected in the application of the constraint.

ACKNOWLEDGMENT

The author wishes to thank Wayne A. Fuller and Rachel M. Harter for sharing a preliminary version of Fuller and Harter (1985), which assisted in the preparation of this paper.

REFERENCES

Efron, B., and Morris, C. (1972), "Limiting the Risk of Bayes and Empirical Bayes Estimators—Part II: The Empirical Bayes Case," *Journal of the American Statistical Association*, **67**, 130–139.

Efron, B., and Morris, C. (1973), "Stein's Estimation Rule and Its Competitors—An Empirical Bayes Approach," *Journal of the American Statistical Association*, **68**, 117–130.

Efron, B., and Morris, C. (1975), "Data Analysis Using Stein's Estimator and Its Generalizations," *Journal of the American Statistical Association*, **70**, 311–319.

Ericksen, E. P. (1973), "A Method of Combining Sample Survey Data and Symptomatic Indicators to Obtain Population Estimates for Local Areas," *Demography*, **10**, 137–160.

Ericksen, E. P. (1974), "A Regression Method for Estimating Population Changes for Local Areas," *Journal of the American Statistical Association*, **69**, 867–875.

Fay, R. E., and Herriot, R. A. (1979), "Estimates of Income for Small Places: An Application of James-Stein Procedures to Census Data," *Journal of the American Statistical Association*, **74**, 269–277.

Fuller, W. A., and Harter, R. M. (1985), "The Multivariate Components of Variance Model for Small Area Estimation," paper presented at the International Symposium on Small Area Statistics, Ottawa, Ontario, Canada, May 22–24, 1985.

Hartley, H. O., and Rao, J. N. K. (1967), "Maximum-Likelihood Estimation for the Mixed Analysis of Variance Model," *Biometrika*, **54**, 93–108.

Harville, D. A. (1976), "Extension of the Gauss-Markov Theorem to Include the Estimation of Random Effects," *Annals of Statistics*, **4**, 384–395.

Harville, D. A. (1977), "Maximum Likelihood Approaches to Variance Component Estimation and to Related Problems," *Journal of the American Statistical Association*, **72**, 320–338.

Hidiroglou, M. A., Fuller, W. A., and Hickman, R. D. (1978), *Super Carp*, 3rd ed., Statistical Laboratory, Iowa State University, Ames, IA.

James, W., and Stein, C. (1961), "Estimation With Quadratic Loss," in *Proceedings of the Fourth Berkeley Symposium on Mathematical Statistics and Probability*, University of California Press, Berkeley, CA, Vol. 1, pp. 361–379.

Morris, C. (1983), "Parametric Empirical Bayes Inference: Theory and Applications," *Journal of the American Statistical Association*, **78**, 47–55.

Purcell, N. J., and Kish, L. (1979), "Estimation for Small Domains," *Biometrics*, **35**, 365–384.

Searle, S. R. (1971), "Topics in Variance Components Estimation," *Biometrics*, **27**, 1–76.

The Multivariate Components of Variance Model for Small Area Estimation

W. A. Fuller and R. M. Harter
Iowa State University

ABSTRACT

The multivariate regression model with components-of-variance error structure is considered. Predictors for small area means are given as the estimated regression mean plus the predictor of the random component associated with the area. Employing expansions in the reciprocals of the degrees of freedom, nearly minimum mean square error predictors and nearly unbiased estimators of the prediction mean square error are constructed. Models with unequal numbers of observations and estimators for finite populations are considered. Results of a Monte Carlo study are presented.

1. INTRODUCTION

The estimation of parameters for small areas has received considerable attention in recent years. A comprehensive review of sample survey research in small area estimation is given by Purcell and Kish (1979). Agencies of the federal government have been significantly involved in research to obtain estimates of such items as population counts, unemployment rates, per capita income, health needs, crop yields, and livestock numbers, for states and local government areas. Acts of the U.S. Congress (e.g., Local Fiscal Assistance Act of 1972 and the National Health Planning and Resources Development Act of 1974) have created a need for accurate small area estimates. Research in this area is illustrated by papers such as Cárdenas, Blanchard, and Craig (1978), DiGaetano et al. (1980), Fay and Herriot (1979), Ericksen (1974), Gonzalez (1973), Gonzalez and Hoza (1978), and Ericksen and Kadane (1985). Särndal (1984) considers the small area problem in the setting with auxiliary information and compares synthetic estimators with design consistent estimators. See also Särndal (1981).

The Fay and Herriot (1979) procedures were based on James–Stein estimation, an approach that is sometimes called empirical Bayes. See Efron and Morris (1973), James and Stein (1961), and Robbins (1955). Morris (1983) gives a description of the empirical Bayes approach and cites additional applications. Dempster, Rubin, and Tsutakawa (1981) give several applications of an extended empirical Bayes model and outline the use of the *EM* algorithm to estimate the variance components.

An early and continued use of the components of variance model for the prediction of random effects is in genetics. See Henderson (1975) and references cited there. Harville (1976, 1985), Kackar and Harville (1984), and Peixoto and Harville (1985) have considered prediction of random effects for such models. Other recent results include those of Reinsel (1984, 1985). Battese and Fuller (1981) used the components of variance model for prediction of crop areas. Also see Fuller and Battese (1973, 1981).

The U.S. Department of Agriculture has been investigating the use of Landsat satellite data to improve its estimates of crop acreages for Crop Reporting Districts and to develop estimates for individual counties. The methodology used in some of these studies is presented in Cárdenas, Blanchard, and Craig (1978), Hanuschak et al. (1979), and Sigman et al. (1978). Also see the special issue of *Communications in Statistics* edited by Chhikara (1984).

2. MODEL AND ESTIMATORS

We consider the model

$$\mathbf{Y}_{ij} = \mathbf{x}_{ij}\mathbf{B} + \mathbf{u}_{ij} \tag{1}$$

where $\mathbf{Y}_{ij}$ is an r-dimensional row vector of observations on the variables for which small area estimates are desired, $\mathbf{x}_{ij}$ is a k-dimensional row vector of auxiliary variables for which the small area totals (or means) are known, $\mathbf{B}$ is a $k \times r$ matrix of coefficients, and $\mathbf{u}_{ij}$ is an r-dimensional row vector of errors. It is assumed that

$$\mathbf{u}_{ij} = \mathbf{v}_i + \mathbf{e}_{ij} \tag{2}$$

$$\mathbf{v}_i \sim NI(\mathbf{0}, \boldsymbol{\Sigma}_{vv}), \qquad \mathbf{e}_{ij} \sim NI(\mathbf{0}, \boldsymbol{\Sigma}_{ee}) \tag{3}$$

where $\mathbf{v}_i$ is independent of $\mathbf{e}_{ij}$ for all i, l, and j. The quantities of interest are the realized means for the individual small areas. Let $\mathbf{y}_i$ denote the realized mean of area i, where

$$\mathbf{y}_i = \bar{\mathbf{x}}_{i(p)}\mathbf{B} + \mathbf{v}_i \tag{4}$$

and $\bar{\mathbf{x}}_{i(p)}$ is the population mean of the $\mathbf{x}_{ij}$ for the ith area, and it is assumed that $\bar{\mathbf{x}}_{i(p)}$ is known.

Assume first that $\boldsymbol{\Sigma}_{vv}$ and $\boldsymbol{\Sigma}_{ee}$ are known. Assume that n_i determinations are made for the ith area and that T areas are observed. There are a total of N observations on the original $\mathbf{Y}_{ij}$-vectors, where $N = \Sigma_{i=1}^{T} n_i$. For the purposes of estimating $\mathbf{B}$, we define the vector of Nr observations on $\mathbf{Y}_{ij}$ and the $Nr \times kr$ matrix $\mathbf{X}$ by

$$\mathbf{Y}' = (\mathbf{Y}_{11}, \mathbf{Y}_{12}, \ldots, \mathbf{Y}_{1, n_1}, \ldots, \mathbf{Y}_{T1}, \ldots, \mathbf{Y}_{T, n_T}),$$

$$\mathbf{X}' = [(\mathbf{I}_r \otimes \mathbf{x}_{11})', (\mathbf{I}_r \otimes \mathbf{x}_{12})', \ldots, (\mathbf{I}_r \otimes \mathbf{x}_{1, n_1})', \ldots, (\mathbf{I}_r \otimes \mathbf{x}_{T, n_T})']$$

where $\otimes$ denotes the Kronecker product. Let

$$\mathbf{V} = \text{block diag}(\mathbf{V}_{11}, \mathbf{V}_{22}, \ldots \mathbf{V}_{TT}) \tag{5}$$

where $\mathbf{V}_{ii} = (\mathbf{J}_{ii} \otimes \boldsymbol{\Sigma}_{vv}) + (\mathbf{I}_{n_i} \otimes \boldsymbol{\Sigma}_{ee})$ and $\mathbf{J}_{ii}$ is the $n_i \times n_i$ matrix with every element equal to one. Let $\mathbf{A} = \text{vec } \mathbf{B}$ denote the column vector of dimension kr obtained by listing the columns of $\mathbf{B}$ one beneath the other beginning with the first column. Then the generalized least squares estimator is

$$\hat{\mathbf{A}} = \text{vec } \hat{\mathbf{B}} = (\mathbf{X}'\mathbf{V}^{-1}\mathbf{X})^{-1}\mathbf{X}'\mathbf{V}^{-1}\mathbf{Y} \tag{6}$$

If $\bar{\mathbf{u}}_{i.}$, the mean of the n_i observations in the ith area, is known, the best predictor of $\mathbf{v}_i$ is the conditional expectation of $\mathbf{v}_i$ given $\bar{\mathbf{u}}_{i.}$. Under the normality assumption,

$$E\{\mathbf{v}_i | \bar{\mathbf{u}}_{i.}\} = \bar{\mathbf{u}}_{i.}\mathbf{G}_i \tag{7}$$

where $\mathbf{G}_i = \mathbf{M}_i^{-1}\boldsymbol{\Sigma}_{vv}$ and $\mathbf{M}_i = \boldsymbol{\Sigma}_{vv} + n_i^{-1}\boldsymbol{\Sigma}_{ee}$. The variance of the prediction error for predictor (7) is

$$E\{(\mathbf{v}_i - \bar{\mathbf{u}}_i \mathbf{G}_i)'(\mathbf{v}_i - \bar{\mathbf{u}}_i \mathbf{G}_i)|\bar{\mathbf{u}}_{i.}\} = \boldsymbol{\Sigma}_{vv} - \boldsymbol{\Sigma}_{vv}(\boldsymbol{\Sigma}_{vv} + n_i^{-1}\boldsymbol{\Sigma}_{ee})^{-1}\boldsymbol{\Sigma}_{vv}$$

Given an estimator of $\mathbf{B}$, one can estimate the $\mathbf{u}_{ij}$, and it is natural to replace the $\mathbf{u}_{ij}$ in (7) with the regression residuals to construct a predictor of $\mathbf{v}_i$. That predictor of $\mathbf{v}_i$ is

$$\hat{\mathbf{v}}_i = \overset{\Delta}{\bar{\mathbf{u}}}_{i.} \mathbf{G}_i \tag{8}$$

where $\overset{\Delta}{\bar{\mathbf{u}}}_{i.} = n_i^{-1}\sum_{j=1}^{n_i}\hat{\mathbf{u}}_{ij}$, and $\hat{\mathbf{u}}_{ij} = \mathbf{Y}_{ij} - \mathbf{X}_{ij}\hat{\mathbf{B}}$. Then, replacing $\mathbf{v}_i$ with the predictor $\hat{\mathbf{v}}_i$ and $\mathbf{B}$ with the estimator $\hat{\mathbf{B}}$ in equation (4), the predictor of the mean for the ith area is

$$\hat{\mathbf{y}}_i = \bar{\mathbf{x}}_{i(p)}\hat{\mathbf{B}} + \hat{\mathbf{v}}_i \tag{9}$$

The covariance matrix of the prediction error for predictor (9) is

$$E\{(\hat{\mathbf{y}}_i - \mathbf{y}_i)'(\hat{\mathbf{y}}_i - \mathbf{y}_i)\} = \boldsymbol{\Sigma}_{vv}(\mathbf{I} - \mathbf{G}_i) + \mathbf{C}_i V\{\hat{\mathbf{A}}\}\mathbf{C}_i' \tag{10}$$

where $V(\hat{\mathbf{A}}) = (\mathbf{X}'\mathbf{V}^{-1}\mathbf{X})^{-1}$, $\mathbf{C}_i = (\mathbf{I}_r \otimes \bar{\mathbf{x}}_{i(p)}) - \mathbf{G}_i'(\mathbf{I}_r \otimes \bar{\mathbf{x}}_{i(s)})$, and $\bar{\mathbf{x}}_{i(s)} = \bar{\mathbf{x}}_{i.}$ is the sample mean of $\mathbf{x}_{ij}$ for the ith area.

In most practical applications it is necessary to estimate the error covariance matrices. There are a number of methods, such as the fitting of constants and restricted maximum likelihood available for estimating the variance components. See Harville (1977). We assume that we construct estimators of $\mathbf{M}_{..} = \boldsymbol{\Sigma}_{vv} + c\boldsymbol{\Sigma}_{ee}$ and of $\boldsymbol{\Sigma}_{ee}$, where c is a known constant. It is assumed that the estimator of $\mathbf{M}_{..}$, denoted by $\hat{\mathbf{M}}_{..}$, is independent of the estimator of $\boldsymbol{\Sigma}_{ee}$. We assume that the unbiased estimator of $\boldsymbol{\Sigma}_{ee}$, denoted by $\hat{\boldsymbol{\Sigma}}_{ee}$, is distributed as a multiple of a Wishart matrix with d_e degrees of freedom. In the balanced case, $\hat{\mathbf{M}}_{..}$ would be the between-area mean square and product matrix of an analysis of variance divided by n, where n is the number of observations in each area, and $\hat{\boldsymbol{\Sigma}}_{ee}$ would be the within-area sample covariance matrix. We assume that a method is used to guarantee that the estimator of $\boldsymbol{\Sigma}_{vv}$, denoted by $\hat{\boldsymbol{\Sigma}}_{vv}$, is positive semidefinite. See Amemiya (1985) and Appendix D.

A reasonable estimator of $\mathbf{G}_i$ is $\hat{\mathbf{G}}_i = \mathbf{I} - \hat{\mathbf{H}}_i$, where $\hat{\mathbf{H}}_i$ is an estimator of $\mathbf{H}_i = (\boldsymbol{\Sigma}_{vv} + n_i^{-1}\boldsymbol{\Sigma}_{ee})^{-1}n_i^{-1}\boldsymbol{\Sigma}_{ee}$. We suggest the nearly unbiased estimator,

$$\hat{\mathbf{H}}_i = [\hat{\mathbf{M}}_i + \hat{\mathbf{K}}_{ii} + (n_i^{-1} - c)^2\hat{\mathbf{W}}_{ii}]^{-1}[n_i^{-1}\hat{\boldsymbol{\Sigma}}_{ee} + (n_i^{-1} - c)n_i^{-1}\hat{\mathbf{W}}_{ii}]$$

where $\hat{\mathbf{M}}_i = \hat{\mathbf{M}}_{..} + (n_i^{-1} - c)\hat{\boldsymbol{\Sigma}}_{ee}$, and $\hat{\mathbf{K}}_{ii}$ and $\hat{\mathbf{W}}_{ii}$ are estimators of

$$\mathbf{K}_{ii} = E\{(\hat{\mathbf{M}}_{..} - \mathbf{M}_{..})\mathbf{M}_i^{-1}(\hat{\mathbf{M}}_{..} - \mathbf{M}_{..})\}$$

and $\mathbf{W}_{ii} = d_e^{-1}[\boldsymbol{\Sigma}_{ee}\mathbf{M}_i^{-1}\boldsymbol{\Sigma}_{ee} + \boldsymbol{\Sigma}_{ee}\,\mathrm{tr}\{\mathbf{M}_i^{-1}\boldsymbol{\Sigma}_{ee}\}]$, respectively. Given regularity conditions, it is shown in the appendices that the estimator $\hat{\mathbf{H}}_i$ has a bias that is $O(d^{-2})$, where d^{-1} is the maximum of T^{-1} and d_e^{-1}. The approximations are developed for sequences of estimators in which both T and d_e increase. The predictor for the mean of the ith area is

$$\tilde{\mathbf{y}}_i = \bar{\mathbf{x}}_{i(p)}\tilde{\mathbf{B}} + \tilde{\mathbf{v}}_i \tag{12}$$

where $\tilde{\mathbf{v}}_i = \bar{\mathbf{u}}_{i.}\hat{\mathbf{G}}_i$, $\bar{\mathbf{u}}_{i.} = n_i^{-1}\sum_{j=1}^{n_j}\tilde{\mathbf{u}}_{ij}$, $\tilde{\mathbf{u}}_{ij} = \mathbf{Y}_{ij} - \mathbf{X}_{ij}\tilde{\mathbf{B}}$,

$$\tilde{\mathbf{A}} = \mathrm{vec}\,\tilde{\mathbf{B}} = (\mathbf{X}'\hat{\mathbf{V}}^{-1}\mathbf{X})^{-1}\mathbf{X}'\hat{\mathbf{V}}^{-1}\mathbf{Y}$$

and $\hat{\mathbf{V}}$ is the covariance matrix (5) with $\hat{\boldsymbol{\Sigma}}_{vv}$ replacing $\boldsymbol{\Sigma}_{vv}$ and $\hat{\boldsymbol{\Sigma}}_{ee}$ replacing $\boldsymbol{\Sigma}_{ee}$.

An approximation to the covariance matrix of the prediction error is

$$\mathbf{V}\{\tilde{\mathbf{y}}_i - \mathbf{y}_i\} = n_i^{-1}\boldsymbol{\Sigma}_{ee}\mathbf{G}_i + \mathbf{C}_i\mathbf{V}\{\hat{\mathbf{A}}\}\mathbf{C}_i' + \mathbf{H}_i'\mathbf{K}_{ii}\mathbf{H}_i + d_e^{-1}\mathbf{M}\,_{..}\mathbf{H}_i\mathbf{M}_i^{-1}\mathbf{H}_i'\mathbf{M}\,_{..}$$
$$+ d_e^{-1}\,\mathrm{tr}\{\mathbf{H}_i\}\mathbf{M}\,_{..}\mathbf{H}_i\mathbf{M}_i^{-1}\mathbf{M}\,_{..} \tag{13}$$

where $\mathbf{C}_i$ is defined in (10). The error in approximation (13) is $O(d^{-2})$. Each of the terms of equation (13) can be associated with a particular source of error. The first term, $n_i^{-1}\boldsymbol{\Sigma}_{ee}\mathbf{G}_i$, is the prediction variance for the situation in which all parameters are known. We call this term the basic variance. The second term, $\mathbf{C}_i\mathbf{V}\{\hat{\mathbf{A}}\}\mathbf{C}_i'$, is the increase in variance that comes from estimating $\mathbf{A} = \mathrm{vec}\,\mathbf{B}$. The third term, $\mathbf{H}_i'\mathbf{K}_{ii}\mathbf{H}_i$, is the contribution to the variance from the estimation of $\mathbf{M}_{..}$. The final two terms of (13) are the contribution to the variance that is due to the estimation of $\boldsymbol{\Sigma}_{ee}$.

On the basis of approximations described in the Appendix, an estimator of the variance of the approximate distribution is

$$\hat{\mathbf{V}}\{\hat{\mathbf{y}}_i - \mathbf{y}_i\} = n_i^{-1}\hat{\boldsymbol{\Sigma}}_{ee} - \hat{\boldsymbol{\Phi}}_{ii} + \mathbf{C}_i\hat{\mathbf{V}}\{\hat{\mathbf{A}}\}\mathbf{C}_i' + \hat{\mathbf{H}}_i'\hat{\mathbf{K}}_{ii}\hat{\mathbf{H}}_i$$
$$+ d_e^{-1}\hat{\mathbf{R}}_i'\hat{\boldsymbol{\Phi}}_{ii}\hat{\mathbf{R}}_i + n_i^{-1}d_e^{-1}\hat{\mathbf{R}}_i'\hat{\boldsymbol{\Sigma}}_{ee}\hat{\mathbf{R}}_i\,\mathrm{tr}\{\hat{\mathbf{H}}_i\} \tag{14}$$

where $\hat{\boldsymbol{\Phi}}_{ii}$ is an estimator of $\boldsymbol{\Phi}_{ii} = n_i^{-2}\boldsymbol{\Sigma}_{ee}\mathbf{M}_i^{-1}\boldsymbol{\Sigma}_{ee}$ defined in equation (C.20) of Appendix C, $\hat{\mathbf{K}}_{ii}$ is an estimator of $\mathbf{K}_{ii}$ defined in equation (D.13) of Appendix D, and

$$\hat{\mathbf{R}}_i = \mathbf{I} - (1 - n_i c)\hat{\mathbf{H}}_i \tag{15}$$

For the balanced case ($n_i \equiv n$ and $\mathbf{x}_{ij} \equiv 1$) an estimator of $\mathbf{H}$ is

$$\tilde{\mathbf{H}} = \alpha_* \hat{\mathbf{M}}_{..}^{-1} n^{-1} \hat{\boldsymbol{\Sigma}}_{ee} \qquad (16)$$

where $\alpha_* = (d_e + r + 1)^{-1} d_e (T - r - 2)(T - 1)^{-1}$ and $\hat{\mathbf{M}}_{..}$ is the between-area mean square matrix divided by n, provided the roots $(\lambda_1, \lambda_2, \ldots, \lambda_r)$ of

$$|\alpha_*^{-1} \hat{\mathbf{M}}_{..} - \lambda n^{-1} \hat{\boldsymbol{\Sigma}}_{ee}| = 0 \qquad (17)$$

satisfy $\lambda_1 \geq \lambda_2 \geq \cdots \geq \lambda_r \geq 1$. A nearly unbiased estimator of the variance for the balanced unknown mean case is

$$\begin{aligned}
\tilde{\mathbf{V}}\{\hat{\mathbf{y}}_i - \mathbf{y}_i\} = {} & n^{-1}\hat{\boldsymbol{\Sigma}}_{ee} - T^{-1}(d_v - r - 1)(d_e + 1)^{-1} d_e n^{-2} \hat{\boldsymbol{\Sigma}}_{ee} \hat{\mathbf{M}}_{..}^{-1} \hat{\boldsymbol{\Sigma}}_{ee} \\
& + [d_e^{-1} + (r + 1)d_v^{-1}]\tilde{\mathbf{H}}' n^{-1} \hat{\boldsymbol{\Sigma}}_{ee} \\
& + 2 d_e^{-1} n^{-1} \hat{\boldsymbol{\Sigma}}_{ee} \operatorname{tr}\{\tilde{\mathbf{H}}\} \qquad (18)
\end{aligned}$$

where $d_v = T - 1$. An unbiased estimator of the variance for the balanced case, containing higher-order terms, is given in Appendix B.

3. MONTE CARLO STUDY

A small Monte Carlo study was conducted on the univariate model to investigate the properties of our approximations for the unequal n_i model. The effect of constructing a nearly unbiased estimator of H_i for the unbalanced model was of particular interest. Each Monte Carlo sample was composed of 10 observations generated to satisfy the model

$$Y_j = \mu + u_j, \qquad u_j \sim NI(0, 0.2 + n_j^{-1})$$

where $\mu = 0$ and the vector of 10 n_j-values is $(1, 1, 2, 2, 3, 4, 5, 7, 10, 15)$. Thus $\sigma_{vv} = 0.2$ and $\sigma_{ee} = 1$. In this study σ_{ee} was assumed to be known and σ_{vv} was estimated.

Two predictors of the area means of the form (12) with $\bar{\mathbf{x}}_{i(p)}\tilde{\mathbf{B}} = \hat{\mu}$ were constructed for each sample. One predictor used the estimator

$$\overset{*}{H}_i = \overset{*}{M}_i{}^{-1} n_i^{-1}$$

where $\overset{*}{M}_i = \max\{n_i^{-1}, \ (1 - 2d_v^{-1})[\hat{M}_{..} + (n_i^{-1} - c)]\}$, $2d_v^{-1} = (\ddot{\sigma}_{vv} + c)^{-2} \hat{V}_{MM}$, and $\hat{V}_{MM}$ and $\ddot{\sigma}_{vv}$ are defined in Appendix A. The estimator $\overset{*}{H}_i$ is roughly equivalent to estimating the two variance components and substituting them into the expression for H_i. The second estimator of H_i

is a form of estimator (11). For σ_{ee} equal to one and known, the estimator has the form

$$\hat{H}_i = \hat{M}_i^{-1} n_i^{-1}$$

where $\hat{M}_i$, and the estimators of μ and $M_{..}$ are defined in Appendix A.

The means of the two estimators of H_i for 100 samples are given in Table 1. The estimator $\hat{H}_i$ is biased downward for all n_i, while $\overset{*}{H}_i$ is biased downward for small n_i and upward for large n_i. At least part of the bias in $\hat{H}_i$ is due to the fact that $\hat{H}_i$ is bounded by one. The estimator $\hat{H}_i$ is uniformly superior to the estimator $\overset{*}{H}_i$. The value of c for this population is 0.21 and the two estimators are nearly identical for $n_i = 5$ $(n_i^{-1} = 0.2)$. The increase in the variance of the prediction error due to estimating the parameters is much smaller for $\hat{y}_i$ than for $\overset{*}{y}_i$ for small and large n_i. For n_i^{-1} close to c the increase in variance is similar for the two estimators.

Properties of the estimator of the prediction variance are given in Table 2. The column headed "Basic" under "Theoretical" is the variance of the predictor obtained using the true parameters. This is the first term of (13). The next column is the large-sample approximation for the increase in variance associated with the estimation of μ and the estimation of H_i. This is the quantity $\mathbf{C}_i \mathbf{V}\{\hat{\mathbf{A}}\} \mathbf{C}_i' + \mathbf{H}_i' \mathbf{K}_{ii} \mathbf{H}_i$ of (13). For the example, the theoretical approximation to the variance of $\hat{\mu}$ is 0.0474 and the theoretical approximation to the variance of $\hat{M}_{..}$ is 0.0421. The Monte Carlo averages of the estimated variances for the basic component are too large because the averages of the estimates $\hat{H}_i$ of Table 1 are too small.

Table 1. Monte Carlo Properties of Alternative Estimators (100 Samples)

n_i	H_i	Mean of Estimates		Estimated Increase in Variance ($\times 100$)	
		$\hat{H}_i$	$\overset{*}{H}_i$	$\hat{y}_i$	$\overset{*}{y}_i$
1	0.833	0.815	0.630	7.99	14.39
1	0.833	0.815	0.630	8.63	14.00
2	0.714	0.686	0.572	6.08	7.20
2	0.714	0.686	0.572	5.12	6.79
3	0.625	0.593	0.529	4.76	5.16
4	0.556	0.523	0.496	4.14	4.24
5	0.500	0.467	0.470	3.10	3.18
7	0.417	0.383	0.430	2.36	2.52
10	0.333	0.296	0.389	1.47	1.75
15	0.250	0.201	0.344	0.67	0.97

Table 2. Monte Carlo Properties of Variance Estimators (100 Samples, All Variances Multiplied by 100)

	Theoretical			Average of Monte Carlo Variance Estimators			Monte Carlo Variance of $\hat{y}_i$
n_i	Basic	Increase from Estimate	Total	Basic	Increase from Estimate	Total	
1	16.67	5.73	22.40	18.51	7.01	25.52	24.98
2	14.29	5.49	19.78	15.70	5.77	21.47	19.89
3	12.50	4.93	17.43	13.56	4.74	18.30	17.26
4	11.11	4.35	15.46	11.93	3.97	15.90	15.25
5	10.00	3.81	13.81	10.66	3.38	14.04	13.10
7	8.33	2.95	11.28	8.83	2.53	11.36	10.69
10	6.67	2.09	8.76	7.06	1.72	8.78	8.14
15	5.00	1.29	6.29	5.33	0.82	6.15	5.67

The estimator of the basic component is $n_i^{-1}(1 - \hat{H}_i)$. The average of the Monte Carlo estimates of the increase in variance due to estimating the parameters is greater than the theoretical increase for small n_i and less than the theoretical increase for large n_i. The average of the estimated variances differs from the observed variance by less than 10% for all n_i.

To study the behavior of the studentized statistics, a sample of 10,000 was created by generating 100 prediction errors for each of the 100 samples. Percentiles of the empirical distribution of

$$\hat{t} = [\hat{V}\{\hat{y}_j\}]^{-1/2}[\mu + v_j - (\hat{\mu} + u_j\hat{G}_j)]$$

are compared with the percentiles of the normal distribution in Table 3. The $\hat{t}$-distribution seems adequately approximated by the normal distri-

Table 3. Percentiles of Studentized Statistics (100 Predictions for 100 Samples)

Sample Size n_i	Percentile			
	0.01	0.05	0.95	0.99
1	−4.06	−2.17	1.94	3.56
2	−3.28	−1.85	1.81	3.00
3	−2.81	−1.80	1.69	2.75
4	−2.65	−1.70	1.68	2.48
5	−2.31	−1.61	1.66	2.47
7	−2.21	−1.59	1.62	2.25
10	−2.25	−1.58	1.57	2.26
15	−2.27	−1.60	1.58	2.21
Normal	−2.33	−1.64	1.64	2.33

bution for large n_i, but the $\hat{t}$-distribution has larger tails than the normal distribution for small n_i.

APPENDIX A: ESTIMATORS IN THE MONTE CARLO STUDY

The statistics constructed for each sample of the Monte Carlo study are given below. The statistics are discussed in Appendices C and D, as well as in the text of the paper.

$$\ddot{\mu} = \sum_{j=1}^{10} \lambda_j Y_j, \qquad \hat{M}_{..} = \left(\sum_{j=1}^{10} n_j b_j \right)^{-1} \sum_{j=1}^{10} n_j (Y_j - \ddot{\mu})^2$$

$$b_i = 1 - 2\lambda_i + \sum_{j=1}^{10} \lambda_j^2$$

$$c = \left(\sum_{j=1}^{10} n_j b_j \right)^{-1} \sum_{i=1}^{10} \left(1 - 2\lambda_i + n_i \sum_{j=1}^{10} \lambda_j^2 n_j^{-1} \right)$$

$$\ddot{\sigma}_{vv} = \max\{0, (T-3)^{-1}(T-1)\hat{M}_{..} - c\}$$

$$\hat{V}_{MM} = 2 \left(\sum_{j=1}^{10} n_j b_j \right)^{-2} \sum_{j=1}^{10} n_j^2 b_j \ddot{M}_j^2$$

$$\ddot{M}_j = \ddot{\sigma}_{vv} + n_j^{-1}, \qquad \hat{M}_i = \max\{n_i^{-1}, \hat{M}_{..} + (n_i^{-1} - c) + \ddot{M}_i^{-1}\hat{V}_{MM}\}$$

$$\hat{y}_i = \hat{\mu} + (Y_i - \hat{\mu})\hat{G}_i, \qquad \hat{H}_i = \hat{M}_i^{-1} n_i^{-1}, \qquad \hat{G}_i = 1 - \hat{H}_i$$

$$\hat{\mu} = \left(\sum_{j=1}^{10} \ddot{M}_j^{-1} \right)^{-1} \sum_{j=1}^{10} \ddot{M}_j^{-1} Y_j, \qquad \hat{V}_{\mu\mu} = \left(\sum_{j=1}^{10} \ddot{M}_j^{-1} \right)^{-1}$$

$$\hat{V}\{\hat{y}_i\} = n_i^{-1} - n_i^{-1}\hat{H}_i + \hat{M}_i^{-1}\hat{V}_{MM}\hat{H}_i^2 + \hat{V}_{\mu\mu}\hat{H}_i^2$$

$$a_j = n_j \text{ if } n_j < 5 \text{ and } a_j = 5 \text{ if } n_j \geq 5, \text{ and } \lambda_j = \left(\sum_{i=1}^{10} a_i \right)^{-1} a_j.$$

APPENDIX B: DERIVATION OF MSE OF PREDICTOR FOR ESTIMATED ERROR COVARIANCES IN A BALANCED DESIGN

We consider the model

$$\mathbf{Y}_{ij} = \mathbf{u} + \mathbf{v}_i + \mathbf{e}_{ij} \tag{B.1}$$

where $i = 1, 2, \ldots, T$ and $j = 1, 2, \ldots, n$. The normal independence assumptions (3) of the text are retained. Let

$$\mathbf{S} = (T-1)^{-1} \sum_{i=1}^{T} (\bar{\mathbf{Y}}_{i.} - \bar{\mathbf{Y}}_{..})'(\bar{\mathbf{Y}}_{i.} - \bar{\mathbf{Y}}_{..}) \tag{B.2}$$

$$\mathbf{W} = [nT(n-1)]^{-1} \sum_{i=1}^{T} \sum_{j=1}^{n} (\mathbf{Y}_{ij} - \bar{\mathbf{Y}}_{i.})'(\mathbf{Y}_{ij} - \bar{\mathbf{Y}}_{i.}) \tag{B.3}$$

$$\bar{\mathbf{Y}}_{i.} = n^{-1} \sum_{j=1}^{n} Y_{ij} \quad \text{and} \quad \bar{\mathbf{Y}}_{..} = T^{-1} \sum_{i=1}^{T} \bar{\mathbf{Y}}_{i.}$$

We study the predictor of a small area mean defined by

$$\hat{\mathbf{y}}_i = \bar{\mathbf{Y}}_{..} + (\bar{\mathbf{Y}}_{i.} - \bar{\mathbf{Y}}_{..})\tilde{\mathbf{G}} = \boldsymbol{\mu} + \bar{\mathbf{v}} + \bar{\mathbf{e}}_{..} + (\bar{\mathbf{u}}_{i.} - \bar{\mathbf{u}}_{..})\tilde{\mathbf{G}} \tag{B.4}$$

where $\tilde{\mathbf{G}} = \mathbf{I} - \alpha \mathbf{S}^{-1}\mathbf{W}$ and α is a fixed real number. The prediction error is

$$\hat{\mathbf{y}}_i - \boldsymbol{\mu} - \mathbf{v}_i = \bar{\mathbf{e}}_{..} + (\bar{\mathbf{v}} - \mathbf{v}_i) + (\bar{\mathbf{u}}_{i.} - \bar{\mathbf{u}}_{..})\tilde{\mathbf{G}} \tag{B.5}$$

and the prediction mean squared error is

$$E\{\bar{\mathbf{e}}'_{..}\,\bar{\mathbf{e}}_{..}\} + E\{[(\mathbf{v}_i - \bar{\mathbf{v}}) - (\bar{\mathbf{u}}_{i.} - \bar{\mathbf{u}}_{..})\tilde{\mathbf{G}}]'[(\mathbf{v}_i - \bar{\mathbf{v}}) - (\bar{\mathbf{u}}_{i.} - \bar{\mathbf{u}}_{..})\tilde{\mathbf{G}}]\} \tag{B.6}$$

where we have used the fact that $\bar{\mathbf{e}}_{..}$ is independent of $\mathbf{v}_i$ and of $(\bar{\mathbf{u}}_{i.} - \bar{\mathbf{u}}_{..})$. For the balanced design, the prediction errors are identically distributed. Therefore (B.6) is equal to the mean squared error averaged over the T areas. The second term of (B.6) summed over the T areas is

$$E\left\{\sum_{i=1}^{T}[(\mathbf{v}_i - \bar{\mathbf{v}})' - \tilde{\mathbf{G}}'(\bar{\mathbf{u}}_{i.} - \bar{\mathbf{u}}_{..})'][(\mathbf{v}_i - \bar{\mathbf{v}}) - (\bar{\mathbf{u}}_{i.} - \bar{\mathbf{u}}_{..})\tilde{\mathbf{G}}]\right\}$$

$$= (T-1)\boldsymbol{\Sigma}_{vv} - E\left\{\tilde{\mathbf{G}}' \sum_{i=1}^{T}(\bar{\mathbf{u}}_{i.} - \bar{\mathbf{u}}_{..})'\mathbf{v}_i\right\} - E\left\{\sum_{i=1}^{T} \mathbf{v}'_i(\bar{\mathbf{u}}_{i.} - \bar{\mathbf{u}}_{..})\tilde{\mathbf{G}}\right\}$$

$$+ E\left\{\tilde{\mathbf{G}}'\left[\sum_{i=1}^{T}(\bar{\mathbf{u}}_{i.} - \bar{\mathbf{u}}_{..})'(\bar{\mathbf{u}}_{i.} - \bar{\mathbf{u}}_{..})\right]\tilde{\mathbf{G}}\right\} \tag{B.7}$$

Under the normal distribution assumptions, $\mathbf{W}$ and $\mathbf{S}$ are independent and

$$E\left\{\left[\sum_{i=1}^{T}(\bar{\mathbf{u}}_{i.}-\bar{\mathbf{u}}_{..})'(\bar{\mathbf{u}}_{i.}-\bar{\mathbf{u}}_{..})\right]^{-1}\sum_{i=1}^{T}(\bar{\mathbf{u}}_{i.}-\bar{\mathbf{u}}_{..})'\mathbf{v}_{i}\right\}$$

$$= (\boldsymbol{\Sigma}_{vv}+n^{-1}\boldsymbol{\Sigma}_{ee})^{-1}\boldsymbol{\Sigma}_{vv}=\mathbf{G} \tag{B.8}$$

Equation (B.8) implies

$$E\left\{\tilde{\mathbf{G}}'\sum_{i=1}^{T}(\bar{\mathbf{u}}_{i.}-\bar{\mathbf{u}}_{..})'\mathbf{v}_{i}\right\}+E\left\{\sum_{i=1}^{T}\mathbf{v}_{i}'(\bar{\mathbf{u}}_{i.}-\bar{\mathbf{u}}_{..})\tilde{\mathbf{G}}\right\}$$

$$= (T-1)2[\boldsymbol{\Sigma}_{vv}-\alpha\boldsymbol{\Sigma}_{vv}\mathbf{H}] \tag{B.9}$$

where $\mathbf{G}=(\boldsymbol{\Sigma}_{vv}+n^{-1}\boldsymbol{\Sigma}_{ee})^{-1}\boldsymbol{\Sigma}_{vv}$ and $\mathbf{H}=n^{-1}(\boldsymbol{\Sigma}_{vv}+n^{-1}\boldsymbol{\Sigma}_{ee})^{-1}\boldsymbol{\Sigma}_{ee}$. The last term in (B.7) can be written

$$E\{(\mathbf{I}-\alpha\mathbf{S}^{-1}\mathbf{W})'(T-1)\mathbf{S}(\mathbf{I}-\alpha\mathbf{S}^{-1}\mathbf{W})\} = (T-1)\mathbf{M}-2(T-1)\alpha n^{-1}\boldsymbol{\Sigma}_{ee}$$
$$+ \alpha^{2}(T-1)^{2}(T-r-2)^{-1}E\{\mathbf{W}\mathbf{M}^{-1}\mathbf{W}\} \tag{B.10}$$

where $\mathbf{M}=(\boldsymbol{\Sigma}_{vv}+n^{-1}\boldsymbol{\Sigma}_{ee})$. Dividing expression (B.7) by T, the expression for the mean square error of $\hat{y}_i$ given in (B.6) is

$$E\{(\hat{\mathbf{y}}_{i}-\mathbf{y}_{i})'(\hat{\mathbf{y}}_{i}-\mathbf{y}_{i})\} = n^{-1}\boldsymbol{\Sigma}_{ee}-2\alpha T^{-1}(T-1)n^{-2}\boldsymbol{\Sigma}_{ee}\mathbf{M}^{-1}\boldsymbol{\Sigma}_{ee}$$
$$+ \alpha^{2}T^{-1}(T-1)^{2}(T-r-2)^{-1}\{n^{-2}\boldsymbol{\Sigma}_{ee}\mathbf{M}^{-1}\boldsymbol{\Sigma}_{ee}$$
$$+ n^{-2}T^{-1}(n-1)^{-1}[\boldsymbol{\Sigma}_{ee}(\operatorname{tr}\{\mathbf{M}^{-1}\boldsymbol{\Sigma}_{ee}\})+\boldsymbol{\Sigma}_{ee}\mathbf{M}^{-1}\boldsymbol{\Sigma}_{ee}]\} \tag{B.11}$$

Expression (B.11) is a generalization of expression (3.10) of Peixoto and Harville (1985), which is a generalization of a result given by Efron and Morris (1973). See also Efron and Morris (1972). If we set

$$\alpha = [d_{v}(d_{e}+2)]^{-1}d_{e}(d_{v}-2) \tag{B.12}$$

where $d_{v}=T-1$ and $d_{e}=T(n-1)$, (B.11) becomes

$$n^{-1}\boldsymbol{\Sigma}_{ee}-T^{-1}(T-1)\alpha n^{-2}\boldsymbol{\Sigma}_{ee}\mathbf{M}^{-1}\boldsymbol{\Sigma}_{ee}$$
$$+ T^{-1}(T-1)\alpha[Tn-T+2]^{-1}n^{-2}[\boldsymbol{\Sigma}_{ee}(\operatorname{tr}\{\mathbf{M}^{-1}\boldsymbol{\Sigma}_{ee}\})-\boldsymbol{\Sigma}_{ee}\mathbf{M}^{-1}\boldsymbol{\Sigma}_{ee}]$$

$$\tag{B.13}$$

This mean squared error is smaller than that obtained with any larger α. If one wishes to minimize $\text{tr}[\Sigma_{ee}^{-1}V\{\hat{y}_i\}]$, the optimal α is

$$\alpha^* = [d_v(d_e + r + 1)]^{-1}(d_v - r - 1)d_e \qquad (B.14)$$

The mean squared error of the estimator with α fixed at α^* is

$$
\begin{aligned}
n^{-1}\Sigma_{ee} &- [T(d_e + r + 1)]^{-1}d_e(d_v - r - 1)n^{-2}\Sigma_{ee}M^{-1}\Sigma_{ee} \\
&+ (d_v - r - 1)d_e[(d_e + r + 1)^2 T]^{-1}n^{-2}[\Sigma_{ee}\,\text{tr}\{M^{-1}\Sigma_{ee}\} \\
&- r\Sigma_{ee}M^{-1}\Sigma_{ee}]
\end{aligned}
\qquad (B.15)
$$

For $r = 1$ expression (B.15) agrees with expression (3.10) of Peixoto and Harville (1985) and the expression for $d_e = \infty$ agrees with results in Efron and Morris (1972, 1973). To estimate the variance we first note that

$$E\{n^2 W\,\text{tr}[WM^{-1}]\} = \Sigma_{ee}\,\text{tr}[\Sigma_{ee}M^{-1}] + 2d_e^{-1}\Sigma_{ee}M^{-1}\Sigma_{ee} \qquad (B.16)$$

and

$$
\begin{aligned}
E\{WS^{-1}W\} &= d_v(d_v - r - 1)^{-1}n^{-2}d_e^{-1}(d_e + 1)\Sigma_{ee}M^{-1}\Sigma_{ee} \\
&+ d_v(d_v - r - 1)^{-1}n^{-2}d_e^{-1}\Sigma_{ee}(\text{tr}\{M^{-1}\Sigma_{ee}\})
\end{aligned}
\qquad (B.17)
$$

Expressions (B.16) and (B.17) lead to a system of equations that can be solved to obtain unbiased estimators of $\Sigma_{ee}M^{-1}\Sigma_{ee}$ and $\Sigma_{ee}\,\text{tr}\{\Sigma_{ee}M^{-1}\}$ as linear functions of $WS^{-1}W$ and $W\,\text{tr}\{WS^{-1}\}$.

If one ignores the second-order terms in d_v^{-1} and in d_e^{-1}, an approximate form for the variance expression (B.15) is

$$
\begin{aligned}
V\{\hat{y}_i - y_i\} &\doteq n^{-1}\Sigma_{ee}G + T^{-1}n^{-2}\Sigma_{ee}M^{-1}\Sigma_{ee} \\
&+ [d_e^{-1} + (r + 1)d_v^{-1}]H'MH + d_e^{-1}n^{-1}\Sigma_{ee}\,\text{tr}\{H\}
\end{aligned}
$$

It follows that an approximately unbiased estimator of the variance is

$$
\begin{aligned}
&n^{-1}\hat{\Sigma}_{ee} - d_v^{-1}(d_v - r - 1)(d_e + 1)^{-1}d_e n^{-2}\hat{\Sigma}_{ee}\hat{M}^{-1}\hat{\Sigma}_{ee}(1 - T^{-1}) \\
&+ [d_e^{-1} + (r + 1)d_v^{-1}]\hat{H}'n^{-1}\hat{\Sigma}_{ee} + 2d_e^{-1}n^{-1}\hat{\Sigma}_{ee}\,\text{tr}\{\hat{H}\}
\end{aligned}
$$

APPENDIX C: APPROXIMATIONS FOR UNEQUAL n_i

In this appendix approximations for the prediction covariance matrix, the estimator of the prediction covariance matrix, and of $H_i = n_i^{-1}M_i^{-1}\Sigma_{ee}$ are developed. We assume that we have two independent matrices estimating the matrices $\mathbf{M}_{..}$ and $\mathbf{\Sigma}_{ee}$. One matrix is a function of the area means of the $\mathbf{u}$ and is denoted by $\hat{\mathbf{M}}_{..}$. The second matrix, denoted by $\hat{\mathbf{\Sigma}}_{ee}$, is an unbiased estimator of $\mathbf{\Sigma}_{ee}$ constructed as the pooled within-mean square obtained from a regression containing dummy variables for the areas. We assume d_e degrees of freedom for the estimator of $\mathbf{\Sigma}_{ee}$.

Let $\hat{\mathbf{u}}_{ij}$ be the residuals from the ordinary least squares regression of $\mathbf{Y}_{ij}$ on $\mathbf{x}_{ij}$, where the regression does not contain area dummies. A class of linear estimators of $\mathbf{M}_{..}$ is

$$\hat{\mathbf{M}}_{..} = \left(\sum_{i=1}^{T} a_i b_i\right)^{-1} \sum_{i=1}^{T} a_i \overset{\Delta}{\hat{\mathbf{u}}}_{i.}' \overset{\Delta}{\hat{\mathbf{u}}}_{i.} \tag{C.1}$$

where $\overset{\Delta}{\hat{\mathbf{u}}}_{i.}$ is the mean of the residuals $\hat{\mathbf{u}}_{ij}$ for the ith area and we assume

$$E\{\overset{\Delta}{\hat{\mathbf{u}}}_{i.}' \overset{\Delta}{\hat{\mathbf{u}}}_{i.}\} = b_i \mathbf{\Sigma}_{vv} + d_i \mathbf{\Sigma}_{ee}$$

with a_i, b_i, and d_i known constants. In the analysis of variance $a_i = n_i$. Under the assumptions,

$$\mathbf{M}_{..} = E\{\hat{\mathbf{M}}_{..}\} = \mathbf{\Sigma}_{vv} + c\mathbf{\Sigma}_{ee} \tag{C.2}$$

where

$$c = \left(\sum_{i=1}^{T} a_i b_i\right)^{-1} \sum_{i=1}^{T} a_i d_i \tag{C.3}$$

Other estimators, such as the restricted maximum likelihood estimator, can be used to estimate $\mathbf{M}_{..}$. An estimator based on estimated optimum weights is given in Appendix D. The assumption retained throughout our analysis is that $\hat{\mathbf{M}}_{..}$ and $\hat{\mathbf{\Sigma}}_{ee}$ are independent.

A simple estimator of $\mathbf{H}_i$ is

$$\dot{\mathbf{H}}_i = \mathbf{I} - \hat{\mathbf{G}}_i = n_i^{-1} \hat{\mathbf{M}}_i^{-1} \hat{\mathbf{\Sigma}}_{ee} \tag{C.4}$$

where $\hat{\mathbf{M}}_i = \hat{\mathbf{M}}_{..} + (n_i^{-1} - c)\hat{\mathbf{\Sigma}}_{ee}$. We will construct an approximately unbiased estimator of $\mathbf{H}_i$ by evaluating the bias in $\dot{\mathbf{H}}_i$. In constructing our approximations we assume that both T and d_e tend to infinity. We have

$$E\{\hat{\mathbf{M}}_i^{-1} - \mathbf{M}_i^{-1}\} = (n_i^{-1} - c)^2 d_e^{-1} \mathbf{M}_i^{-1} [\mathbf{\Sigma}_{ee} \mathbf{M}_i^{-1} \mathbf{\Sigma}_{ee} + \mathbf{\Sigma}_{ee}(\mathrm{tr}\{\mathbf{M}_i^{-1}\mathbf{\Sigma}_{ee}\})]\mathbf{M}_i^{-1}$$
$$+ \mathbf{M}_i^{-1} \mathbf{K}_{ii} \mathbf{M}_i^{-1} + O(T^{-2}) \tag{C.5}$$

where

$$\mathbf{K}_{ii} = E\{(\hat{\mathbf{M}}_{..} - \mathbf{M}_{..})\mathbf{M}_i^{-1}(\hat{\mathbf{M}}_{..} - \mathbf{M}_{..})\} \tag{C.6}$$

If $\hat{\mathbf{M}}_{..}$ is an estimator in the class (C.1), then

$$\mathbf{K}_{ii} \doteq \left(\sum_{j=1}^T a_j b_j\right)^{-2} \sum_{j=1}^T a_j^2 b_j [\mathbf{M}_j \mathbf{M}_i^{-1} \mathbf{M}_j + \mathbf{M}_j \operatorname{tr}\{\mathbf{M}_i^{-1}\mathbf{M}_j\}] \tag{C.7}$$

In the approximation (C.7) we use b_j rather than b_j^2 in the numerator because the $\hat{\mathbf{u}}_{i.}$ are not independent. Expression (C.7) reduces to the correct expression for the balanced unknown mean model. Also

$$E\{\mathbf{M}_i^{-1}(n_i^{-1} - c)(\hat{\boldsymbol{\Sigma}}_{ee} - \boldsymbol{\Sigma}_{ee})\mathbf{M}_i^{-1}n_i^{-1}(\hat{\boldsymbol{\Sigma}}_{ee} - \boldsymbol{\Sigma}_{ee})\}$$
$$= \mathbf{M}_i^{-1}(n_i^{-1} - c)n_i^{-1}d_e^{-1}[\boldsymbol{\Sigma}_{ee}(\operatorname{tr}\{\mathbf{M}_i^{-1}\boldsymbol{\Sigma}_{ee}\}) + \boldsymbol{\Sigma}_{ee}\mathbf{M}_i^{-1}\boldsymbol{\Sigma}_{ee}] \tag{C.8}$$

Therefore,

$$E\{n_i^{-1}\hat{\mathbf{M}}_i^{-1}\hat{\boldsymbol{\Sigma}}_{ee}\} \doteq \mathbf{H}_i + \mathbf{M}_i^{-1}\mathbf{K}_{ii}\mathbf{H}_i + (1 - n_i c)^2 d_e^{-1}\mathbf{H}_i^3$$
$$+ (1 - n_i c)^2 d_e^{-1} \operatorname{tr}\{\mathbf{H}_i\}\mathbf{H}_i^2 - (1 - n_i c)d_e^{-1}\mathbf{H}_i^2$$
$$- (1 - n_i c)d_e^{-1}\mathbf{H}_i \operatorname{tr}\{\mathbf{H}_i\} \tag{C.9}$$

It follows from (C.5) and (C.8) that a nearly unbiased estimator of $\mathbf{H}_i$ is

$$\hat{\mathbf{H}}_i = [\hat{\mathbf{M}}_i + \hat{\mathbf{K}}_{ii} + (n_i^{-1} - c)^2\hat{\mathbf{W}}_{ii}]^{-1}[n_i^{-1}\hat{\boldsymbol{\Sigma}}_{ee} + (n_i^{-1} - c)n_i^{-1}\hat{\mathbf{W}}_{ii}] \tag{C.10}$$

where $\hat{\mathbf{K}}_{ii}$ is an estimator of $\mathbf{K}_{ii}$ and

$$\hat{\mathbf{W}}_{ii} = d_e^{-1}[\hat{\boldsymbol{\Sigma}}_{ee}\hat{\mathbf{M}}_i^{-1}\hat{\boldsymbol{\Sigma}}_{ee} + \hat{\boldsymbol{\Sigma}}_{ee} \operatorname{tr}\{\hat{\mathbf{M}}_i^{-1}\hat{\boldsymbol{\Sigma}}_{ee}\}] \tag{C.11}$$

See appendix D, equation (D.13), for an estimator of $\mathbf{K}_{ii}$.

An estimator of $\mathbf{H}_i$ that recognizes the effect of estimating $\boldsymbol{\Sigma}_{ee}$ is

$$\tilde{\mathbf{H}}_i = d_e[d_e + r^{-1}(1 + r) \operatorname{tr}\{\hat{\mathbf{M}}_{..}\hat{\mathbf{M}}_i^{-1}\hat{\mathbf{M}}_{..}\hat{\mathbf{M}}_i^{-1}\}]^{-1}\hat{\mathbf{H}}_i \tag{C.12}$$

For moderately large d_e it may be reasonable to use the simpler estimator $\hat{\mathbf{H}}_i$.

We now return to the estimation of the mean for the ith area. The estimation error is

$$\tilde{\mathbf{y}}_i - \mathbf{y}_i = \bar{\mathbf{x}}_{i(p)}(\tilde{\mathbf{B}} - \mathbf{B}) + \hat{\mathbf{u}}_{i.}\tilde{\mathbf{G}}_i - \mathbf{v}_i \tag{C.13}$$

where $\tilde{\mathbf{G}}_i = \mathbf{I} - \tilde{\mathbf{H}}_i$, $\tilde{\mathbf{y}}_i = \bar{\mathbf{x}}_{i(p)}\tilde{\mathbf{B}} + \overset{\triangle}{\mathbf{u}}_{i\cdot}\tilde{\mathbf{G}}_i$, and $\tilde{\mathbf{A}} = (\mathbf{X}'\hat{\mathbf{V}}^{-1}\mathbf{X})^{-1}\mathbf{X}'\hat{\mathbf{V}}^{-1}\mathbf{Y}$. The difference between $\hat{\mathbf{B}}$, the estimator of $\mathbf{B}$ based on the true covariance matrix, and $\tilde{\mathbf{B}}$, the estimator of $\mathbf{B}$ based on the estimated covariance matrix, is of relatively small order, and we ignore that difference in constructing an approximation to the covariance matrix of the estimator. The error in (C.13) can be written

$$\tilde{\mathbf{y}}_i - \mathbf{y}_i = \bar{\mathbf{u}}_{i\cdot}\mathbf{G}_i - \mathbf{v}_i + (\hat{\mathbf{A}} - \mathbf{A})'\mathbf{C}_i' + \bar{\mathbf{u}}_{i\cdot}(\tilde{\mathbf{G}}_i - \mathbf{G}_i) + O_p(N^{-1}) \quad (C.14)$$

where $\mathbf{C}_i$ is defined in (10) and $\hat{\mathbf{A}} = \text{vec}\,\hat{\mathbf{B}}$. The first three terms on the right of (C.14) are uncorrelated. See Kackar and Harville (1984). It follows that

$$\hat{\mathbf{H}}_i - \mathbf{H}_i \doteq \mathbf{M}_i^{-1}(\hat{\mathbf{\Sigma}}_{ee} - \mathbf{\Sigma}_{ee})n_i^{-1}\mathbf{R}_i - \mathbf{M}_i^{-1}(\hat{\mathbf{M}}_{\cdot\cdot} - \mathbf{M}_{\cdot\cdot})\mathbf{H}_i \quad (C.15)$$

where $\mathbf{R}_i = \mathbf{I} - (1 - n_i c)\mathbf{H}_i = \mathbf{M}_i^{-1}\mathbf{M}_{\cdot\cdot}$. Also

$$E\{(\tilde{\mathbf{G}}_i - \mathbf{G}_i)'\bar{\mathbf{u}}_{i\cdot}'\bar{\mathbf{u}}_{i\cdot}(\tilde{\mathbf{G}}_i - \mathbf{G}_i)\} \doteq \mathbf{H}_i'\mathbf{K}_{ii}\mathbf{H}_i + d_e^{-1}\mathbf{R}_i'\mathbf{H}_i'\mathbf{M}_i\mathbf{H}_i\mathbf{R}_i$$
$$+ d_e^{-1}\,\text{tr}\{\mathbf{H}_i\}\mathbf{R}_i'\mathbf{M}_i\mathbf{H}_i\mathbf{R}_i \quad (C.16)$$

Therefore, the approximate covariance matrix of $\tilde{\mathbf{y}}_i - \mathbf{y}_i$ is

$$V\{\tilde{\mathbf{y}}_i - \mathbf{y}_i\} \doteq n_i^{-1}\mathbf{\Sigma}_{ee} - \mathbf{\Phi}_{ii} + \mathbf{C}_i V\{\hat{\mathbf{A}}\}\mathbf{C}_i' + \mathbf{H}_i'\mathbf{K}_{ii}\mathbf{H}_i$$
$$+ d_e^{-1}\mathbf{R}_i'\mathbf{\Phi}_{ii}\mathbf{R}_i + n_i^{-1}d_e^{-1}\,\text{tr}\{\mathbf{H}_i\}\mathbf{R}_i'\mathbf{\Sigma}_{ee}\mathbf{R}_i \quad (C.17)$$

where $\mathbf{K}_{ii}$ is defined in (C.6), $\mathbf{C}_i$ is defined in (10) of the text,

$$\mathbf{\Phi}_{ii} = n_i^{-2}\mathbf{\Sigma}_{ee}\mathbf{M}_i^{-1}\mathbf{\Sigma}_{ee}$$

and $\mathbf{R}_i$ is defined in (C.15).

To estimate the covariance matrix of $\tilde{\mathbf{y}}_i - \mathbf{y}_i$, we must construct an estimator of $n_i^{-2}\mathbf{\Sigma}_{ee}\mathbf{M}_i^{-1}\mathbf{\Sigma}_{ee} = \mathbf{\Phi}_{ii}$. Expression (C.10) provides a starting point and we define

$$\ddot{\mathbf{\Phi}}_{ii} = n_i^{-2}[\hat{\mathbf{\Sigma}}_{ee} + (n_i^{-1} - c)\hat{\mathbf{W}}_{ii}][\hat{\mathbf{M}}_i + \hat{\mathbf{K}}_{ii}$$
$$+ (n_i^{-1} - c)^2\hat{\mathbf{W}}_{ii}]^{-1}[\hat{\mathbf{\Sigma}}_{ee} + (n_i^{-1} - c)\hat{\mathbf{W}}_{ii}] \quad (C.18)$$

where $\hat{\mathbf{W}}_{ii}$ is defined in (C.11). By our usual expansions

$$E\{\ddot{\mathbf{\Phi}}_{ii}\} \doteq \mathbf{\Phi}_{ii} + d_e^{-1}[\mathbf{\Phi}_{ii} + n_i^{-2}\mathbf{\Sigma}_{ee}\,\text{tr}\{\mathbf{M}_i^{-1}\mathbf{\Sigma}_{ee}\}] \quad (C.19)$$

It follows that a nearly unbiased estimator of $\mathbf{\Phi}_{ii}$ is

$$\hat{\mathbf{\Phi}}_{ii} = (d_e + 1)^{-1} d_e \ddot{\mathbf{\Phi}}_{ii} - d_e^{-1} n_i^{-1} \hat{\mathbf{\Sigma}}_{ee} \operatorname{tr}\{\tilde{\mathbf{H}}_i\} \qquad (\text{C.20})$$

On the basis of (C.17) and (C.20) an estimator of $\mathbf{V}\{\bar{\mathbf{y}}_i - \mathbf{y}_i\}$ that has bias of order $d_e^{-2} + T^{-1} d_e^{-1}$ is

$$\hat{\mathbf{V}}\{\bar{\mathbf{y}}_i - \mathbf{y}_i\} = n_i^{-1} \hat{\mathbf{\Sigma}}_{ee} - \hat{\mathbf{\Phi}}_{ii} + \hat{\mathbf{C}}_i \hat{\mathbf{V}}\{\hat{\mathbf{A}}\} \hat{\mathbf{C}}_i' + \tilde{\mathbf{H}}_i' \hat{\mathbf{K}}_{ii} \tilde{\mathbf{H}}_i$$
$$+ d_e^{-1} \tilde{\mathbf{R}}_i' \hat{\mathbf{\Phi}}_{ii} \tilde{\mathbf{R}}_i + n_i^{-1} d_e^{-1} \operatorname{tr}\{\tilde{\mathbf{H}}_i\} \tilde{\mathbf{R}}_i' \hat{\mathbf{\Sigma}}_{ee} \tilde{\mathbf{R}}_i \qquad (\text{C.21})$$

where $\hat{\mathbf{C}}_i' = (\mathbf{I}_r \otimes \bar{\mathbf{x}}_{i(p)}') - (\mathbf{I}_r \otimes \bar{\mathbf{x}}_{i(s)}') \hat{\mathbf{G}}_i$, $\tilde{\mathbf{R}}_i = \mathbf{I} - (1 - n_i c) \tilde{\mathbf{H}}_i$, and an estimator of $\mathbf{K}_{ii}$ is defined in Appendix D.

APPENDIX D: NOTES ON THE CONSTRUCTION OF ESTIMATORS

We assume that we first construct an estimator of $\mathbf{M}_{..}$ as a member of the class (C.1) and that we have an estimator of $\mathbf{\Sigma}_{ee}$ that is independent of the estimator of $\mathbf{M}_{..}$. Let $\ddot{\mathbf{Q}}$ be the matrix such that

$$\ddot{\mathbf{Q}}' \hat{\mathbf{\Sigma}}_{ee} \ddot{\mathbf{Q}} = \mathbf{I} \qquad (\text{D.1})$$

$$(T - k - r - 1)^{-1} (T - k) \ddot{\mathbf{Q}}' \hat{\mathbf{M}}_{..} \ddot{\mathbf{Q}} = \ddot{\mathbf{\Lambda}} \qquad (\text{D.2})$$

where $\ddot{\mathbf{\Lambda}} = \operatorname{diag}(\ddot{\lambda}_{11}, \ddot{\lambda}_{22}, \ldots, \ddot{\lambda}_{rr})$. Let

$$\ddot{\lambda}_{ii*} = \max(c, \ddot{\lambda}_{ii}) \qquad (\text{D.3})$$

$$\ddot{\mathbf{\Lambda}}_{**} = \operatorname{diag}(\ddot{\lambda}_{11*}, \ddot{\lambda}_{22*}, \ldots, \ddot{\lambda}_{rr*}) \qquad (\text{D.4})$$

$$\ddot{\mathbf{\Sigma}}_{ff} = \ddot{\mathbf{\Lambda}}_{**} - c\mathbf{I} \qquad (\text{D.5})$$

where c is defined in (C.3).

The multiplier used for $\hat{\mathbf{M}}_{..}$ is the inverse of the approximation to the multiplier one would apply to $\hat{\mathbf{M}}^{-1}$ to create a nearly unbiased estimator of $\mathbf{M}^{-1}$ for the balanced model. The $\ddot{\mathbf{Q}}$ is an estimator of the transformation that reduces the $\bar{\mathbf{u}}_{i.}$ to a vector of uncorrelated random variables. If $\mathbf{Q}$ is the corresponding population matrix, then

$$E\{\mathbf{Q}' \bar{\mathbf{u}}_{i.} \bar{\mathbf{u}}_{i.} \mathbf{Q}\} = \mathbf{\Sigma}_{ff} + n_i^{-1} \mathbf{I} \qquad (\text{D.6})$$

where $\mathbf{Q'\Sigma}_{ee}\mathbf{Q} = \mathbf{I}$ and

$$\Sigma_{ff} = \mathbf{Q'\Sigma}_{vv}\mathbf{Q} = \text{diag}(\lambda_1 - c, \lambda_2 - c, \ldots, \lambda_r - c)$$
$$= \text{diag}(\sigma_{ff11}, \sigma_{ff22}, \ldots, \sigma_{ffrr})$$

One may choose to compute improved estimators of $\mathbf{M}_{..}$ at this point. Because the sample covariances for a normal vector with diagonal covariance matrix are uncorrelated, improved estimators of the elements of the transformed $\mathbf{M}_{..}$ are given by

$$\hat{\lambda}_{\xi\xi jl} = \left(\sum_{i=1}^{T} a_{i(jl)} b_{i(jl)} \right)^{-1} \sum_{i=1}^{T} a_{i(jl)} \overset{\Delta}{\xi}_{ij} \overset{\Delta}{\xi}_{il}' \tag{D.7}$$

where $\overset{\Delta}{\xi}_{i.}' = \ddot{\mathbf{Q}}'\mathbf{u}_{i.}'$, $\overset{\Delta}{\xi}_{i.} = (\overset{\Delta}{\xi}_{i1}, \overset{\Delta}{\xi}_{i2}, \ldots, \overset{\Delta}{\xi}_{ir})$,

$$a_{i(jl)} = [(\ddot{\sigma}_{ffjj} + n_i^{-1})(\ddot{\sigma}_{ffll} + n_i^{-1})(1 + \delta_{ij})]^{-1}$$
$$E\{\overset{\Delta}{\xi}_{ij}\overset{\Delta}{\xi}_{il}'\} = b_{i(jl)}\sigma_{ffjl} + d_{i(jl)}$$

$b_{i(jl)}$ and $d_{i(jl)}$ are known constants, $\ddot{\sigma}_{ffjj} = \ddot{\lambda}_{jj*} - c$, $\ddot{\lambda}_{jj*}$ is defined in (D.3) and δ_{ij} is Kronecker's delta. The use of second round estimators of a new $\mathbf{M}_{..}$ will require computation of second round estimators of $\ddot{\mathbf{Q}}$ and $\ddot{\mathbf{\Sigma}}_{ff}$.

The estimators of the variances of the estimated elements of the matrix $\boldsymbol{\Lambda}$ are

$$\hat{V}\{\hat{\lambda}_{jl}\} = \left(\sum_{i=1}^{T} a_{i(jl)} b_{i(jl)} \right)^{-2} \left(\sum_{i=1}^{T} a_{i(jl)}^2 b_{i(jl)} (\ddot{\sigma}_{ffjj} + n_i^{-1})(\ddot{\sigma}_{ffll} + n_i^{-1}) \right.$$
$$\left. \times (1 + \delta_{lj}) \right) \tag{D.8}$$

and the estimated covariance between $\hat{\lambda}_{jl}$ and $\hat{\lambda}_{ks}$ is zero if $jl \neq ks$. Now

$$\mathbf{K}_{ii} = E\{\mathbf{Q}^{-1}{}'\mathbf{Q}'(\hat{\mathbf{M}}_{..} - \mathbf{M}_{..})\mathbf{Q}(\mathbf{Q'M}_i\mathbf{Q})^{-1}\mathbf{Q}'(\hat{\mathbf{M}}_{..} - \mathbf{M}_{..})\mathbf{QQ}^{-1}\} \tag{D.9}$$

where $\mathbf{Q'M}_i\mathbf{Q} = \boldsymbol{\Lambda} + (n_i^{-1} - c)\mathbf{I} = \boldsymbol{\Sigma}_{ff} + n_i^{-1}\mathbf{I}$, and

$$\mathbf{Q}'(\hat{\mathbf{M}}_{..} - \mathbf{M}_{..})\mathbf{Q} = \hat{\boldsymbol{\Lambda}} - \boldsymbol{\Lambda}$$

By (D.8), the matrix

$$E\{(\hat{\Lambda} - \Lambda)(\Sigma_{ff} + n_i^{-1}I)^{-1}(\hat{\Lambda} - \Lambda)\} \tag{D.10}$$

is a diagonal matrix with llth element given by

$$\sum_{j=1}^{r} (\sigma_{ffjj} + n_i^{-1})^{-1}V\{\hat{\lambda}_{jl}\} \tag{D.11}$$

The elements of (D.11) can be estimated by

$$\hat{\omega}_{ffll} = \sum_{j=1}^{r} (\ddot{\sigma}_{ffjj} + n_i^{-1})^{-1}\hat{V}\{\hat{\lambda}_{jl}\} \tag{D.12}$$

where $\ddot{\Sigma}_{ff}$ is defined in (D.5) and $\hat{V}\{\hat{\lambda}_{jl}\}$ is defined in (D.8). Then an estimator of $\mathbf{K}_{ii}$ is

$$\hat{\mathbf{K}}_{ii} = \ddot{\mathbf{Q}}^{-1\prime}\hat{\Omega}_{ff}\ddot{\mathbf{Q}}^{-1}, \tag{D.13}$$

where $\ddot{\mathbf{Q}}$ is defined by (D.1) and (D.2), and

$$\hat{\Omega}_{ff} = \text{diag}\{\hat{\omega}_{ff11}, \hat{\omega}_{ff22}, \ldots, \hat{\omega}_{ffrr}\}$$

The estimator (D.13) can be used in the estimators (C.10), (C.18), and (C.21). The use of (D.13) as it stands in (C.10) will not produce the optimum estimator for the balanced case. A modification that we have used is to replace $\ddot{\sigma}_{ffjj}$ in (D.8) and (D.12) with

$$\tilde{\sigma}_{ffjj} = \max\{0, (T - r - k - 1)^{-1}(T - k)\ddot{\lambda}_{jj*} - c\} \tag{D.14}$$

where $T - k$ is the raw degrees of freedom for the estimator of $\mathbf{M}_{..}$. Equation (D.14) produces the optimum estimator in the balanced unknown means case ($k = 1$). Limited evidence suggests that the estimator (D.14) also produces better estimators of the variance of the predictor when used in (C.18).

APPENDIX E: SMALL AREA ESTIMATION FOR FINITE POPULATIONS

We assume that we know the population mean, denoted by $\bar{\mathbf{x}}_{i(p)}$, for a vector of auxiliary variables for the ith area. We assume that the population of N_i elements for the ith area was generated by the superpopulation model

$$\mathbf{Y}_{ij} = \mathbf{x}_{ij}\mathbf{B} + \mathbf{u}_{ij} \tag{E.1}$$

where the $\mathbf{u}_{ij}$ satisfy our components of variance model.

Assume that n_i observations are available for the ith area. Then the mean of those n_i observations is estimated by the observed mean and the mean for the unobserved $N_i - n_i$ elements is estimated by the model estimator defined in (12),

$$\bar{\mathbf{y}}_{i(p-s)} = \bar{\mathbf{x}}_{i(p-s)}\tilde{\mathbf{B}} + \tilde{\mathbf{v}}_i \tag{E.2}$$

where $\tilde{\mathbf{v}}_i = \bar{\mathbf{u}}_{i.}\tilde{\mathbf{G}}_i$ and $\bar{\mathbf{x}}_{i(N-n)} = \bar{\mathbf{x}}_{i(p-s)}$ is the mean for the $N_i - n_i$ unobserved units in the ith area. Therefore our estimator of the finite population mean is

$$\tilde{\mathbf{y}}_{i(p)} = f_i \bar{\mathbf{y}}_{i(s)} + (1 - f_i)[\bar{\mathbf{x}}_{i(p-s)}\tilde{\mathbf{B}} + \tilde{\mathbf{v}}_i] \tag{E.3}$$

where $f_i = N_i^{-1}n_i$. It follows that

$$\tilde{\mathbf{y}}_{i(p)} - \bar{\mathbf{y}}_{i(p)} = (1 - f_i)[\bar{\mathbf{x}}_{i(p-s)}(\tilde{\mathbf{B}} - \mathbf{B}) + \tilde{\mathbf{v}}_i - \mathbf{v}_i - \bar{\mathbf{e}}_{i(p-s)}]$$

where $\bar{\mathbf{y}}_{i(p)} = \bar{\mathbf{x}}_{i(p)}\mathbf{B} + \mathbf{v}_i + \bar{\mathbf{e}}_{i(p)}$ is the finite population mean. We have

$$\mathbf{V}\{\tilde{\mathbf{y}}_i - \bar{\mathbf{y}}_{i(p)}\} \doteq (1 - f_1)^2[\mathbf{C}_{if}\mathbf{V}\{\hat{\mathbf{A}}\}\mathbf{C}'_{if} + \mathbf{V}\{\bar{\mathbf{u}}_{i.}\mathbf{G}_i - \mathbf{v}_i$$
$$+ \bar{\mathbf{u}}_{i.}(\tilde{\mathbf{G}}_i - \mathbf{G}_i)\} + (N_i - n_i)^{-1}\boldsymbol{\Sigma}_{ee}] \tag{E.4}$$

where $\mathbf{C}_{if} = (\mathbf{I}_r \otimes \bar{\mathbf{x}}_{i(p-s)}) - (\mathbf{G}'_i \otimes \bar{\mathbf{x}}_{i(s)})$, and

$$\mathbf{V}\{\bar{\mathbf{u}}_{i.}\mathbf{G}_i - \mathbf{v}_i + \bar{\mathbf{u}}_{i.}(\tilde{\mathbf{G}}_i - \mathbf{G}_i)\} \doteq n_i^{-1}\boldsymbol{\Sigma}_{ee}\mathbf{G}_i + \mathbf{H}'_i\mathbf{K}_{ii}\mathbf{H}_i$$
$$+ d_e^{-1}\mathbf{M}_{..}(\mathbf{H}_i\mathbf{M}_i^{-1}\mathbf{H}'_i + \mathbf{H}_i\mathbf{M}_i^{-1}\,\mathrm{tr}\{\mathbf{H}_i\})\mathbf{M}_{..}$$

If there are no observations in an area ($n_i = 0$), then the estimator is $\tilde{\mathbf{y}}_{i(p)} = \bar{\mathbf{x}}_{i(p)}\hat{\mathbf{B}}$ and the approximate variance is

$$(\mathbf{I}_r \otimes \bar{\mathbf{x}}_{i(p)})\mathbf{V}\{\hat{\mathbf{A}}\}(\mathbf{I}_r \otimes \bar{\mathbf{x}}_{i(p)})' + \boldsymbol{\Sigma}_{vv} + N_i^{-1}\boldsymbol{\Sigma}_{ee} \tag{E.5}$$

REFERENCES

Amemiya, Y. (1985), "What Should be Done When an Estimated Between-Group Covariance Matrix is Not Nonnegative Definite?" *The American Statistician*, **39**, 112–117.

Battese, G. E., and Fuller, W. A. (1981), "Prediction of County Crop Areas Using Survey and Satellite Data," *Proceedings of the Survey Research Methods Section of the American Statistical Association*, pp. 500–505.

Cárdenas M., Blanchard, M. M., and Craig, M. E. (1978), *On the Development of Small Area Estimators Using LANDSAT Data as Auxiliary Information*, Economics, Statistics, and Cooperatives Service, U.S. Department of Agriculture, Washington, DC.

Chhikara, R. S., Ed. (1984), "Crop Surveys Using Satellite Data," *Communications in Statistics–Theory and Methods* **13** (Special Issue).

Dempster, A. P., Rubin, D. B. and Tsutakawa, R. K. (1981), "Estimation in Covariance Components Models," *Journal of the American Statistical Association*, **76**, 341–353.

DiGaetano, R., MacKenzie, E., Waksberg, J., and Yaffe, R. (1980), "Synthetic Estimates for Local Areas from the Health Interview Survey," *1980 Proceedings of the Survey Research Methods Section of the American Statistical Association*, pp. 46–55.

Efron, B., and Morris, C. (1972), "Empirical Bayes on Vector Observations: An Extension of Stein's Method," *Biometrika*, **59**, 335–347.

Efron, B., and Morris, C. (1973), "Stein's Estimation Rule and Its Competitors–An Empirical Bayes Approach," *Journal of the American Statistical Association*, **68**, 117–130.

Ericksen, E. P. (1974), "A Regression Method for Estimating Population Changes of Local Areas," *Journal of the American Statistical Association*, **69**, 867–875.

Ericksen, E. P. and Kadane, J. B., (1985), "Estimating the Population in a Census Year: 1980 and Beyond," *Journal of the American Statistical Associatin*, **80**, 98–109.

Fay, R. E., and Herriot, R. (1979), "Estimates of Income for Small Places: An Application of James-Stein Procedures to Census Data," *Journal of the American Statistical Association*, **74**, 269–277.

Fuller, W. A., and Battese, G. E. (1973), "Transformations of Estimation of Linear Models with Nested-Error Structure," *Journal of the American Statistical Association*, **68**, 626–632.

Fuller, W. A., and Battese, G. E. (1981), "Regression Estimation for Small Areas," in *Rural America in Passage: Statistics for Policy* (D. M. Gilford, G. L. Nelson, and L. Ingram, eds.), National Academy Press, Washington, DC.

Gonzalez, M. E. (1973), "Use and Evaluation of Synthetic Estimates," *1973 Proceedings of the Social Statistics Section*, American Statistical Association, pp. 33–36.

Gonzalez, M. E., and Hoza, C. (1978), "Small Area Estimation with Application to Unemployment and Housing Estimates," *Journal of the American Statistical Association*, **73**, 7–15.

Hanuschak, G., Sigman, R., Craig, M., Ozga, M., Luebbe, R., Cook, P., Kleweno, D., and Miller, C. (1979), *Obtaining Timely Crop Area Estimates Using Ground-Gathered and LANDSAT Data* (Technical Bulletin No. 1609), Economics, Statistics, and Cooperatives Service, U.S. Department of Agriculture, Washington, DC.

Harter, R. M. (1983), *Small Area Estimation Using Nested-Error Models and Auxiliary Data* (unpublished Ph.D. thesis), Iowa State University, Ames, IA.

Harville, D. A. (1976), "Extension of the Gauss–Markov Theorem to Include the Estimation of Random Effects," *The Annals of Statistics*, **4**, 384–395.

Harville, D. A. (1977), "Maximum Likelihood Approaches to Variance Component Estimation and to Related Problems," *Journal of the American Statistical Association*, **72**, 320–338.

Harville, D. A. (1979), "Some Useful Representations for Constrained Mixed-Model Estimation," *Journal of the American Statistical Association*, **74**, 200–206.

Harville, D. A. (1985), "Decomposition of Prediction Error," *Journal of the American Statistical Association*, **80**, 132–138.

Henderson, C. R. (1975), "Best Linear Unbiased Estimation and Prediction Under a Selection Model," *Biometrics*, **31**, 423–447.

Hidiroglou, M. A., Fuller, W. A., and Hickman, R. D. (1980), *SUPER CARP*, 6th ed., Survey Section, Statistical Laboratory, Iowa State University, Ames, IA.

Hidiroglou, M. A., Morry, M., Dagum, E. B., and Rao, J. N. K. (1984), "Evaluation of Alternative Small Area Estimators Using Administrative Records," *Proceedings of the Survey Research Methods Section of the American Statistical Association*, pp. 307–313.

James, W., and Stein, C. (1961), "Estimation with Quadratic Loss," *Proceedings of the Fourth Berkeley Symposium of Mathematical Statistics and Probability*, University of California Press, Berkeley, CA, Vol. 1, pp. 361–379.

Kackar, R. N., and Harville, D. A. (1984), "Approximations for Standard Errors of Estimators of Fixed and Random Effects in Mixed Linear Models," *Journal of the American Statistical Association*, **79**, 853–862.

Morris, C. N. (1983), "Parametric Empirical Bayes Inference: Theory and Applications," *Journal of the American Statistical Association*, **78**, 47–54.

Peixoto, J. L. (1982), *Estimation of Random Effects in the Balanced One-Way Classification*, (Unpublished Ph.D. thesis), Iowa State University, Ames, IA.

Peixoto, J. L., and Harville, D. A. (1986), "Comparisons of Alternative Predictors Under the Balanced One-Way Random Model," *Journal of the American Statistical Association*, **81**, 431–436.

Purcell, N. J., and Kish, L. (1979), "Estimation for Small Domains," *Biometrics*, **35**, 365–384.

Reinsel, G. C. (1984), "Estimation and Prediction in a Multivariate Random Effects Generalized Linear Model," *Journal of the American Statistical Association*, **79**, 406–414.

Reinsel, G. C. (1985), "Mean Squared Error Properties of Empirical Bayes Estimators in a Multivariate Random Effects General Linear Model," *Journal of the American Statistical Association*, **80**, 642–650.

Robbins, H. (1955), "An Empirical Bayes Approach to Statistics," *Proceedings of the Third Berkeley Symposium on Mathematics, Statistics and Probability*, University of California Press, Berkeley, CA, Vol. 1, pp. 157–163.

Särndal, C. E. (1981), "Frameworks for Inference in Survey Sampling with Applications to Small Area Estimation and Adjustment for Nonresponse," *Bulletin of the International Statistical Institute*, **49**(1), 494–513.

Särndal, C. E. (1984), "Design–Consistent versus Model-Dependent Estimation for Small Domains," *Journal of the American Statistical Association*, **79**, 624–631.

Searle, S. R. (1971), *Linear Models*, Wiley, New York.

Sigman, R. S., Hanuschak, G. A., Craig, M. E., Cook, P. W., and Cardenas, M. (1978), "The Use of Regression Estimation with LANDSAT and Probability Ground Sample Data," *1978 Proceedings of the Section on Survey Research Methods*, American Statistical Association, pp. 165–168.

Walker, G., and Sigman, R. (1982), *The Use of LANDSAT for County Estimates of Crop Areas: Evaluation of the Huddleston-Ray and the Battese-Fuller Estimators* (SRS Staff Report No. AGES-820909), U.S. Department of Agriculture, Washington, DC.

Bayes and Empirical Bayes Approaches to Small Area Estimation

T. W. F. Stroud
Queen's University at Kingston

ABSTRACT

Let $a = 1, 2, \ldots, A$ represent subpopulations (areas) and let μ_a be the mean over subpopulation a of a variable Y. Suppose that for each subpopulation we have a sample of size n_a on Y and a value x_a of an auxiliary variable X, which is known to be related to the variable Y. Specifically, assume that, for $a = 1, \ldots, A$, the parameter μ_a has a normal prior distribution with the mean $\alpha + \beta x_a$ and variance τ^2, where α, β, and τ^2 are unknown. The form of the distribution of Y within subpopulations is assumed known (e.g., Gaussian or Poisson). For the Bayesian approach, α, β, and τ^2 are given diffuse priors. Formulas are obtained for the posterior mean and variance of each μ_a for the case of equal n_a. For the empirical Bayes solution, α, β, and τ^2 are estimated. This method is illustrated with data from a sample survey of Queen's students.

1. INTRODUCTION AND SUMMARY

In simultaneously estimating the mean $\bar{Y}_a$ of a variable in each of a large number of small areas, one method of attack is to find a compromise between the sample mean of an area $\bar{y}_a$ and an estimator $\tilde{y}_a$ based on a regression on one or more covariates, for example, Fay and Herriot (1979). The disadvantage of $\bar{y}_a$ as an estimator is that it ignores the covariates, whereas $\tilde{y}_a$ ignores the fact that each area has idiosyncratic features not accounted for by the covariates. One way of combining the

124

two sources of information is to use a hierarchical Bayesian (two-stage Bayesian) approach, where the prior distribution of the unknown area means contains one or more hyperparameters that are subject to a relatively diffuse prior distribution. In this article, this hierarchical Bayesian approach is developed for the case of equal sample sizes. We derive exact formulas for the posterior mean and posterior variance of the parameters of interest, the area population means. Normal distributions are assumed both for the population of measurements within each area and for the prior distribution of the area means about a linear function of the covariate.

For simplicity of presentation, this article involves but one covariate; however, the extension to two or more covariates is straightforward and poses no methodological problems.

For the case of unequal sample sizes taken from a large number of areas, we recommended empirical Bayes techniques that replace the second stage of the Bayesian formulation by point estimation of the hyperparameters. One way of doing this is described in this article; another approach is given by Fay and Herriot (1979).

If the sample sizes are unequal but the number of areas is small, the empirical Bayes approach is overly simplistic in that it ignores the uncertainty concerning the variance in the means about the linear function of the covariate. For this case, the hierarchical Bayesian (hereinafter referred to simply as "Bayesian") approach is more reliable, even though more complicated. The analysis is not presented in this article, but is possible as an extension of this article using the technique of Stroud (1984), that is, numerical integration, possible with series expansions.

In this presentation, it is assumed that for each area there is but one applicable value of the covariate, rather than a covariate value corresponding to each separate unit within the area. In situations where the latter exists, the method can still be applied, using the sample mean of the covariate over the area as the applicable covariate value, or using the population mean if it is available. We do not cover the case where one wishes to adjust for a discrepancy between sample and population values of the covariate. A different model is required for this case; however, the Bayesian approach of this paper could still be applied.

Section 2 contains the first-stage Bayesian analysis (for general sample sizes) conditional on the variance parameters, Section 3 completes the hierarchical Bayesian analysis for the case of equal sample sizes, and Section 4 presents an illustrative example. Section 5 briefly describes an empirical Bayes approach. Finally, there is an Appendix containing the heavier integrations needed in Section 3.

2. CONDITIONAL POSTERIOR DISTRIBUTIONS
GIVEN THE VARIANCES

Let y_{ai} be the value of the variable of interest y measured on the ith sampled unit within the ath area $(a = 1, \ldots, A; i = 1, \ldots, n_a)$. According to the assumed model, the y_{ai}, given $\mu_1, \ldots, \mu_A$ and σ^2, are independently normally distributed with $E(y_{ai}) = \mu_a$ and $V(y_{ai}) = \sigma^2$. Given the values of α, β, τ^2, and of the known auxiliary variable x_a $(a = 1, \ldots, A)$, the μ_a have a normal prior distribution such that $E(\mu_a) = \alpha + \beta x_a$, $V(\mu_a) = \tau^2$, and the μ_a are independent. We assume that α and β are jointly distributed according to a diffuse (flat) prior in the region of the plane where the μ_a are concentrated; this may be approximated either by a uniform distribution over the plane or by bivariate normal distribution, the inverse of whose covariance matrix tends to zero. At this stage, both σ^2 and τ^2 are assumed known.

The resulting posterior distribution for $(\mu_1, \ldots, \mu_A)$ may be obtained by straightforward Bayesian analysis or, more conveniently, from the formulas in Lindley and Smith (1972). It is normal with covariance matrix.

$$D = \{(\sigma^{-2}N + \tau^{-2}I) - \tau^{-2}C(C'C)^{-1}C'\}^{-1}$$

where $N = \mathrm{diag}(n_1, \ldots, n_A)$, I is the identity matrix, and

$$C' = \begin{bmatrix} 1 & 1 & \cdots & 1 \\ x_1 & x_2 & \cdots & x_A \end{bmatrix}$$

Using a familiar matrix inversion identity [see, e.g., Lindley and Smith (1972, formula 10)], this can be rewritten as

$$D = (\sigma^{-2}N + \tau^{-2}I)^{-1} + (\sigma^{-2}N + \tau^{-2}I)^{-1}CEC'(\sigma^{-2}N + \tau^{-2}I)^{-1}$$

where

$$E^{-1} = \tau^4 \sigma^{-2} \begin{bmatrix} \sum \Psi_a & \sum \Psi_a x_a \\ \sum \Psi_a x_a & \sum \Psi_a x_a^2 \end{bmatrix}$$

and $\Psi_a = 1/[(\tau^2/\sigma^2) + (1/n_a)]$. The mean of this posterior distribution is $\sigma^{-2}DN\bar{y}$, where $\bar{y}$ is the vector of sample means $\bar{y} = (\bar{y}_1, \ldots, \bar{y}_A)'$. Thus the conditional posterior variances and covariances of the μ_a may be written

$$V(\mu_a \mid y; \sigma^2, \tau^2) = \frac{1}{\dfrac{n_a}{\sigma^2} + \dfrac{1}{\tau^2}} + \frac{e_{11} + 2e_{12}x_a + e_{22}x_a^2}{\left(\dfrac{n_a}{\sigma^2} + \dfrac{1}{\tau^2}\right)^2} \tag{1}$$

$$\text{cov}(\mu_a, \mu_b \mid y; \sigma^2, \tau^2) = \frac{e_{11} + e_{12}(x_a + x_b) + e_{22}x_a x_b}{\left(\dfrac{n_a}{\sigma^2} + \dfrac{1}{\tau^2}\right)\left(\dfrac{n_b}{\sigma^2} + \dfrac{1}{\tau^2}\right)}$$

where $y \equiv (y_{11}, \ldots, y_{An_A})'$ and e_{11}, e_{12}, e_{22} are the components of the symmetric matrix E.

By further straightforward manipulation, the conditional posterior mean of μ_a may be written

$$E(\mu_a \mid y; \sigma^2, \tau^2) = \left(1 - \frac{\Psi_a}{n_a}\right)\bar{y}_a + \frac{\Psi_a}{n_a}\tilde{y}_a \tag{2}$$

where

$$\tilde{y}_a = \tau^4 \sigma^{-2} \sum_{b=1}^{A} [e_{11} + e_{12}(x_a + x_b) + e_{22}x_a x_b]\Psi_b \bar{y}_b \tag{3}$$

Thus the posterior mean shrinks the sample mean $\bar{y}_a$ toward $\tilde{y}_a$ with a shrinking weight of Ψ_a/n_a; where $\tilde{y}_a$ is a particular weighted average of $\bar{y}_1, \ldots, \bar{y}_A$.

3. POSTERIOR MEANS AND VARIANCES IN THE FULL BAYESIAN ANALYSIS WITH EQUAL SAMPLE SIZES

Since in practice σ^2 and, especially, τ^2 are unknown, a complete Bayesian analysis requires a prior distribution on σ^2 and τ^2. In the case of equal sample sizes, this can be done in a way that is straightforward and practical. In this subsection, we assume that $n_1 = \cdots n_A \equiv n$. This implies that $\Psi_1 = \Psi_2 = \cdots = \Psi_A \equiv \Psi$.

Instead of σ^2 and τ^2, it is convenient to let the hyperparameters be λ and Ψ, where $\lambda = 1/\sigma^2$ and $\Psi = 1/[(\tau^2/\sigma^2) + (1/n)]$. Note that $0 < \sigma^2 < \infty$ and $0 \leq \tau^2 < \infty$ imply $0 < \lambda < \infty$ and $0 < \Psi \leq n$. For λ, we assume a scaled chi-square distribution with ν_0 degrees of freedom and scale factor m_0. We assume that Ψ is independent of λ and that it belongs to a family of priors proposed by Strawderman (1971) for one-way analysis-of-variance model. Since in the full posterior analysis the likelihood of $\mu = (\mu_1, \ldots, \mu_A)'$ also involves the hyperparameters α and β, we include α

and β in our formulation here and then integrate them out when appropriate. Thus the assumed joint prior on the hyperparameters is

$$p(\alpha, \beta, \lambda, \Psi) \propto \Psi^{-r} \lambda^{(\nu_0/2)-1} e^{-\lambda \nu_0 m_0/2},$$
$$-\infty < \alpha, \beta < \infty, \lambda > 0, 0 < \Psi \le n \qquad (4)$$

In the Strawderman family of proper priors, $0 < r < 1$. We shall include the limiting improper case $r = 1$ as well as the limiting improper case of the scaled chi-squared $\nu_0 = 0$ (Jeffreys prior).

We note that the form of the transformation defining Ψ implies that the prior structure on σ^2, τ^2 depends on n. This type of prior has been advocated by a number of authors, for example, Box and Tiao (1968, 1973) and Strawderman (1971), because of the resulting mathematical simplicity, although others have criticized it from a coherence point of view. We take the position that there are many possible diffuse priors and, although they lead to slightly different results with small samples, the choice of one over the other is a toss-up, and one might as well use one that yields computable results.

Note that the problem we are studying is an extension of the Bayesian solution of one-way components of variance. If β is forced to be zero, the problem reduces to components of variance. We are allowing the μ_a to vary about a linear function of x_a instead of about a constant grand mean. If the prior structure used by Box and Tiao (1968, 1973) in their Bayesian solution of balanced one-way components of variance is extended to the problem we are considering, the result is equivalent to (4) in the limiting improper case $r = 1$, $\nu_0 = 0$.

Although the complete posterior distribution $p(\mu, \alpha, \beta, \lambda, \Psi \mid y)$ is obtainable in principle, it would involve a great deal of writing and is not needed for the task at hand. For the purposes of this paper, we desire formulas for $E(\mu_a \mid y)$ and $V(\mu_a \mid y)$, since these would give us a point estimate of the mean for any desired area, together with a measure of uncertainty of that estimate.

Taking expectations of both sides of (2), and noting $\Psi_a = \Psi$, $n_a = n$, we get

$$E(\mu_a \mid y) = \bar{y}_a + \frac{E(\Psi \mid y)}{n} (\tilde{y}_a - \bar{y}_a) \qquad (5)$$

where for the case of equal sample size (3) reduces to

$$\tilde{y}_a = \bar{\bar{y}} + \hat{\beta}(x_a - \bar{x})$$

where $\hat{\beta} = \Sigma(x_a - \bar{x})(\bar{y}_a - \bar{\bar{y}}) / \Sigma(x_a - \bar{x})^2$ and $\bar{\bar{y}} = (\Sigma \bar{y}_a)/A$.

The corresponding method for obtaining the variance is to use

$$V(\mu_a \mid y) = E\{V(\mu_a \mid y; \sigma^2, \tau^2)\} + V\{E(\mu_a \mid y; \sigma^2, \tau^2)\} \qquad (6)$$

where for equal sample sizes (1) reduces to

$$V(\mu_a \mid y; \sigma^2, \tau^2) = \frac{1}{\lambda n} - \frac{\Psi}{\lambda n^2}\left(1 - \frac{1}{A} - \frac{(x_a - \bar{x})^2}{S}\right),$$

where $S = \Sigma(x_a - \bar{x})^2$. Substituting this expression into (6) yields

$$V(\mu_a \mid y) = \frac{1}{n}E(1/\lambda \mid y) - \frac{1}{n^2}E(\Psi/\lambda \mid y)\left(1 - \frac{1}{A} - \frac{(x_a - \bar{x})^2}{S}\right)$$

$$+ \frac{(\bar{y}_a - \bar{y}_a)^2}{n^2}[E(\Psi^2 \mid y) - \{E(\Psi \mid y)\}^2] \qquad (7)$$

Formulas for the expectations of $E(\Psi \mid y)$, $E(\Psi^2 \mid y)$, $E(\lambda^{-1} \mid y)$, and $E(\Psi\lambda^{-1} \mid y)$ are derived by integration over the joint posterior distribution of Ψ and λ, and appear in the Appendix as formulas (A.3) through (A.5).

4. AN ILLUSTRATIVE APPLICATION TO SURVEY DATA

We illustrate the use of the results of Section 3 by means of a small data set taken from a survey whose purpose was to teach sample surveys to students at Queen's University. The questionnaire item, administered to university students, was "how many trips home do you estimate you will have taken by the end of the academic year?" The area a in this analysis is the municipality of the student's home address. In the present example we consider Arts and Science students from the municipalities of Belleville, Calgary, Montreal (including other municipalities on the island), Oakville, Ottawa (including Nepean, Gloucester and Kanata), Sault Ste. Marie, Metropolitan Toronto, and Vancouver. This is the set of municipalities other than Kingston with three or more sampled Arts and Science students. The samples analyzed here consist of three students from each area; where the survey data contained more than three, three were randomly sampled.

It was decided that the auxiliary variable should be a function of road distance between the municipality and Kingston. To explore which function, a graph of log mean number of trips versus log distance was plotted for a larger data set (all faculties, 15 municipalities instead of 8, all available data points arranged into the mean number of trips). Since the plot of the 15 points exhibited a reasonably modest scatter about linearity, the least-squares value of the slope was calculated; it was -0.494. Thus it seemed reasonable that the x-variable should be the $-\frac{1}{2}$ power of the distance from the municipality from Kingston. Although, theoretically, living at an infinite distance from Kingston should imply zero trips and hence the line fitting $\bar{y}_a$ to x_a might reasonably be forced to pass through the origin, it was decided to retain the intercept parameter α, which is present throughout the entirety of this paper, as a possible compensation for mild departures from the model.

For the data being analyzed, $A = 8$ and $n = 3$, Table 1 gives the values of distance $= 1/x^2$, x, $\bar{y}$, and the fitted value $\hat{y}$ for each of the eight municipalities. The grand mean was the point $(0.0543, 4.417)$ and the slope was $\hat{\beta} = 98.4$. The within-group sum of squares was $SSW = 509.33$. The between-group sum of squares (ignoring distances) was $SSB = 254.5$, yielding an F ratio of 1.142, indicating that with this size data set it is not at all clear that number of trips home is related to municipality. However, when the distance-related variable x is brought in, SSB is decomposed into a regression sum of squares $n\hat{\beta}^2 S = 214.3$ on one degree of freedom and a departure from linearity sum of squares $nSSE = 40.2$ on six degrees of freedom. Consequently, the variable x accounts for an F of 6.73 measured against SSW (degrees of freedom $= 1,16$) or 8.58 measured against $SSW + nSSE$ (degrees of freedom $= 1,22$); so that the relationship between x and mean number of trips is substantial.

The desired estimate for the population means of the eight municipalities lies somewhere between the respective values in the last two columns. If the mean were a perfect linear function of the reciprocal square root of the distance, then the last column would give the best

Table 1. Mean Number of Trips and Its Linear Fit to (Distance)$^{-1/2}$ for 8 Municipalities

Municipality a	Distance $=$ x_a^{-2} (km)	x_a	Average $\bar{y}_a$	Fitted $\hat{y}_a$
Belleville	79	0.1125	12.33	10.14
Calgary	3434	0.0171	1.33	0.75
Montreal	298	0.0579	4.00	4.77
Oakville	290	0.0587	4.33	4.85
Ottawa	167	0.0774	4.67	6.69
Sault Ste. Marie	882	0.0337	3.67	2.39
Toronto	257	0.0624	4.00	5.21
Vancouver	4492	0.0149	1.00	0.54

estimate. The second last column would be best if number of trips home were thought to be unrelated to distance. The specific compromise we propose is given by formula (5).

Using the uninformative prior $\nu_0 = 0$, $r = 1$, formulas (A.3) through (A.5) yield $E(\Psi \mid y) = 2 \cdot 1684$, $E(\Psi^2 \mid y) = 5 \cdot 1076$, $E\{(1/\lambda) \mid y\} = 18 \cdot 867$ and $E\{(\Psi/\lambda) \mid y\} = 41 \cdot 175$. These formulas are substituted into (5) and (6) to yield the values of the posterior means $E(\mu_a \mid y)$ and the posterior variances $V(\mu_a \mid y)$ for the eight municipalities. Table 2 shows these values, together with $E(\mu_a \mid y) \pm 2\sqrt{V(\mu_a \mid y)}$, which may be regarded as approximate 95% limits.

We may wish to compare the precision as embodied in $V(\mu_a \mid y)$ with the estimated precision from classical analysis, in particular, with a one-way ANOVA model and with a linear regression model. For the one-way ANOVA model, the fitted values for the eight municipalities are just the sample means; the variance of each of these estimates, based on an assumed common within-group variance σ_w^2, is given by $s_w^2/3 = 10.611$, which is considerably greater than any of the values of $V(\mu_a \mid y)$. For the regression model, the variance of the estimates $\bar{\bar{y}} + \hat{\beta}(x_a - \bar{x})$, given by $V(\bar{y}) + (x_a - \bar{x})^2 V(\hat{\beta})$ ranges from 1.055 for Montreal to 4.860 for Belleville, in most cases somewhat less than $V(\mu_a \mid y)$. The variance of the regression estimate $\bar{y} + \hat{\beta}(x_a - \bar{x})$ is based on a MSE of 24.979 with 22 degrees of freedom. According to the analysis of the third paragraph of this section, the difference between groups is very well explained by the explanatory variable x.

Under the assumption that there is no explanatory power in the groups outside the variable x, the classical analysis seems to do better. However, everything is based on only 24 observations. The Bayesian model presented here incorporates the prior assumption that the μ_a are not exactly equal to $\alpha + \beta x_a$, but fluctuate about this value.

Table 2. Posterior Means, Variances, Approximate 95% Limits

Municipality a	$E(\mu_a \mid y)$	$V(\mu_a \mid y)$	$E - 2\sqrt{V}$	$E + 2\sqrt{V}$
Belleville	10.75	4.387	7.15	14.35
Calgary	0.91	3.147	−2.64	4.46
Montreal	4.56	2.294	1.53	7.59
Oakville	4.71	2.298	1.68	7.74
Ottawa	6.13	2.614	2.90	9.36
Sault Ste. Marie	2.74	2.552	−0.45	5.93
Toronto	4.87	2.326	1.82	7.92
Vancouver	0.67	3.248	−2.93	4.27

5.   EMPIRICAL BAYES APPROACH

When the sample sizes are unequal, which is the usual situation, the full Bayesian analysis becomes more cumbersome. It is possible here to define Ψ as the harmonic mean of the Ψ_a, and then to proceed according to the method used in Stroud (1984), Ngai (1982), and Nebebe (1984). If the number of areas is large, one must choose between doing a large number of numerical integrations and utilizing series expansions. In the latter case, there may be a convergence problem. It is more straightforward to use the empirical Bayes approach, that is, to replace the joint posterior distribution of σ^2 and τ^2 (or transformations of same) by point estimates. The problem here is in estimating τ^2. If A is small, any estimate will be imprecise; but if A is large, the information about τ^2 will be sufficiently great that the discrepancy in the results between using an estimate of τ^2 and using a posterior distribution becomes a second-order problem.

There have been a number of articles on the empirical Bayes approach in the recent literature, for example, Fay and Herriot (1979), Morris (1983), and Casella (1985); we shall utilize the formulation of Dempster, Rubin, and Tsutakawa (1981), specifically, the formulation they refer to as LR, which stands for "(maximum) likelihood, random (parameters)." We write the model described in Section 2 as

$$y_{ai} = \alpha + \beta x_a + \gamma_a + e_{ai}, \qquad i = 1, \ldots, n_a, a = 1, \ldots, A$$

where e_{ai} is the usual-style error term and $\gamma_a = \mu_a - (\alpha + \beta x_a)$. Thus the γ_a are independently distributed as $N(0, \tau^2)$. Now the model fits the class of models described in Dempster et al. (1981), where their β_1 is our $(\alpha, \beta)'$, and their β_2 is our $(\gamma_1, \ldots, \gamma_A)'$. Their matrix Σ is specified by $\Sigma_{12} = 0$, $\Sigma_{22} = \tau^2 I$, and $\Sigma_{11}^{-1} \to 0$. The first column in the matrix they denote by X has 1 everywhere, the second column has x_a, while columns 3 through $A + 2$, respectively, are dummy variables denoting membership in one of the A areas. The number of rows in X is $\Sigma_a n_a$. The unknown parameters σ^2 and τ^2 are estimated by maximum likelihood using the EM algorithm (Dempster, Laird, and Rubin, 1977) as described in Dempster et al. (1981, pp. 343–344). When the resulting estimates are substituted into (1) and (2), one obtains approximations to posterior variance and posterior mean, respectively, of the μ_a. As noted in Dempster et al. (1981, p. 344), formula (1) so employed will tend to underestimate the posterior variance, owing mainly to the fact that the posterior uncertainty about τ^2 is being ignored. Further work is required to show the extent of this inaccuracy.

APPENDIX. POSTERIOR DENSITY OF λ AND Ψ, AND CERTAIN POSTERIOR MOMENTS

We begin with (4), together with $p(y \mid \alpha, \beta, \lambda, \Psi)$, which is the marginal distribution of y with respect to μ, conditional on the hyperparameters. This marginal distribution [see, for example, Lindley and Smith (1972, formula 6)] is given by the density

$$p(y \mid \alpha, \beta, \lambda, \Psi) = \frac{\lambda^{nA/2}}{(2\pi)^{nA/2}\left|I_{nA} + \left(\dfrac{1}{\Psi} - \dfrac{1}{n}\right)J_n \otimes I_A\right|^{1/2}}$$

$$\times \exp\left\{-\frac{\lambda}{2}(y - \alpha e - \beta x)'\left[I_{nA} + \left(\frac{1}{\Psi} - \frac{1}{n}\right)J_n \otimes I_A\right]^{-1}(y - \alpha e - \beta x)\right\}$$

where e is a vector of ones of length nA, x is the vector of length nA corresponding to y, and I_t and J_t represent identity matrix and matrix of ones, respectively, of order t. Multiplication of this likelihood of $(\alpha, \beta, \lambda, \Psi)$ by the prior (4) and simplification yields, by Bayes' rule,

$$p(\alpha, \beta, \lambda, \Psi \mid y) \propto \frac{\Psi^{-r}\lambda^{(\nu_0 + nA - 2)/2} e^{-\lambda(\nu_0 m_0 + SSW)/2}}{(n/\Psi)^{A/2}}$$

$$\times \exp\left\{-\frac{\lambda\Psi}{2}\Sigma_{a=1}^{A}[\alpha - (\bar{y}_a - \beta x_a)]^2\right\} \tag{A.1}$$

where $SSW = \Sigma\Sigma(y_{ai} - \bar{y}_a)^2$, the sum of squares within areas.

The joint posterior density of the hyperparameters is now determined, up to a multiplicative constant.

We shall integrate out α and β to obtain the joint posterior density of (λ, Ψ), from which we shall obtain $E(\Psi \mid y)$, $E(\Psi^2 \mid y)$, $E(1/\lambda \mid y)$, and $E(\Psi/\lambda \mid y)$. It will be seen that these four quantities, together with formulas (5) and (6), determine $E(\mu_a \mid y)$ and $V(\mu_a \mid y)$.

We note from (A.1) that the factor involving α is a normal density. After completing the square and integrating out α, we get

$$p(\beta, \lambda, \Psi \mid y) \propto \Psi^{(A - 2r - 1)/2}\lambda^{(\nu_0 + nA - 3)/2} e^{-\lambda(\nu_0 m_0 + SSW)/2}$$

$$\times \exp\left\{-\frac{\lambda\Psi}{2}\sum_{a=1}^{A}[(x_a - \bar{x})\beta - (\bar{y}_a - \bar{y}_\cdot)]^2\right\}$$

where $\bar{y}_\cdot$ is the grand mean, in this case $(\Sigma y_a)/A$.

After integrating out β, the result may be written

$$p(\lambda, \Psi \mid y) = k^* \Psi^{(A/2)-r-1} \lambda^{(\nu_0 + nA - 4)/2} e^{-\lambda(\nu_0 m_0 + SSW + \Psi SSE)/2}$$

where SSE equals $\Sigma(\bar{y}_a - \bar{y}.)^2 - [\Sigma(x_a - \bar{x})(\bar{y}_a - \bar{y}.)]^2 / \Sigma(x_a - \bar{x})^2$, the residual sum of squares in the lines regression of the $\bar{y}_a$ on the x_a, and k^* is the normalizing constant in this bivariate density. If λ is integrated out, the result is

$$p(\Psi \mid y) = \frac{k^* \Gamma\{(\nu_0 + nA - 2)/2\}}{\{(\nu_0 m_0 + SSW + \Psi SSE)/2\}^{(\nu_0 + nA - 2)/2}}, \qquad 0 < \Psi \le n$$

To evaluate k^*, we need the value of

$$\int_0^n \frac{\Psi^{(A/2)-r-1} \, d\Psi}{(\nu_0 m_0 + SSW + \Psi SSE)^{(\nu_0 + nA - 2)/2}} \tag{A.2}$$

This may be transformed into a beta-type integral. Rewrite (A.2) as

$$(\nu_0 m_0 + SSW)^{-(\nu_0 + nA - 2)/2} \int_0^n \frac{\Psi^{(A/2)-r-1} \, d\Psi}{[1 + (SSE)\Psi/(\nu_0 m_0 + SSW)]^{(\nu_0 + nA - 2)/2}}$$

and making the substitution $y = 1 - 1/[1 + \Psi SSE/(\nu_0 m_0 + SSW)]$ we find that

$$\int_0^n \frac{\Psi^{(A/2)-r-1} \, d\Psi}{(\nu_0 m_0 + SSW + \Psi SSE)^{(\nu_0 + nA - 2)/2}}$$

$$= \left(\frac{\nu_0 m_0 + SSW}{SSE}\right)^{A/2-r} \int_0^w y^{A/2-r-1} (1-y)^{(\nu_0 + \nu_1)/2 + r - 2} \, dy$$

$$= \frac{B\left(\frac{A}{2} - r, \frac{\nu_0 + \nu_1}{2} + r - 1\right) I_w\left(\frac{A}{2} - r, \frac{\nu_0 + \nu_1}{2} + r - 1\right)}{(SSE)^{A/2-r}(\nu_0 m_0 + SSW)^{[(\nu_0 + \nu_1)/2] + r - 1}}$$

where $\nu_1 = (n-1)A$, the within-group degrees of freedom, and $w = nSSE/(\nu_0 m_0 + SSW + nSSE)$.

From this we see that

$$k^* = \frac{(SSE)^{A/2-r}(\nu_0 m_0 + SSW)^{[(\nu_0+\nu_1)/2]+r-1}}{2^{(\nu_0+nA-1)/2}\Gamma((\nu_0 + nA - 2)/2)B\left(\dfrac{A}{2} - r, \dfrac{\nu_0 + \nu_1}{2} + r - 1\right)}$$

$$\times \left[I_w\left(\frac{A}{2} - r, \frac{\nu_0 + \nu_1}{2} + r - 1\right)\right]^{-1}$$

and that hence

$$p(\Psi \mid y) = \frac{(SSE)^{A/2-r}(\nu_0 m_0 + SSW)^{[(\nu_0+\nu_1)/2]+r-1}}{B\left(\dfrac{A}{2} - r, \dfrac{\nu_0 + \nu_1}{2} + r - 1\right)I_w\left(\dfrac{A}{2} - r, \dfrac{\nu_0 + \nu_1}{2} + r - 1\right)}$$

$$\times \frac{\Psi^{(A/2)-r-1}}{(\nu_0 m_0 + SSW + \Psi SSE)^{(\nu_0+nA-2)/2}}, \qquad 0 < \Psi < n$$

We see from this that an arbitrary moment of Ψ, say $E(\Psi^k \mid y)$, may be found by repeating the integration of (A.2) with r replaced by $r - k$. If this is carried out, the result is

$$E(\Psi^k \mid y) = \left(\frac{\nu_0 m_0 + SSW}{SSE}\right)^k \frac{B\left(\dfrac{\nu_2}{2} + k + 1 - r, \dfrac{\nu_0 + \nu_1}{2} - k + r - 1\right)}{B\left(\dfrac{\nu_2}{2} + 1 - r, \dfrac{\nu_0 + \nu_1}{2} + r - 1\right)}$$

$$\times \frac{I_w\left(\dfrac{\nu_2}{2} + k + 1 - r, \dfrac{\nu_0 + \nu_1}{2} - k + r - 1\right)}{I_w\left(\dfrac{\nu_2}{2} + 1 - r, \dfrac{\nu_0 + \nu_1}{2} + r - 1\right)} \qquad \text{(A.3)}$$

where $\nu_2 = A - 2$, the residual degrees of freedom after fitting the regression on x to the A means. Formula (A.3) is used at the end of Section 3 for the cases $k = 1$ and $k = 2$.

It is useful to compare (A.3) with formula (7.2.27) of Box and Tiao (1973). For the one-way analysis-of-variance model, that formula gives

$$E\left(\frac{\Psi}{n}\right)^k = \frac{B\left(\dfrac{\nu_2}{2} + k, \dfrac{\nu_1}{2} - k\right)I_w\left(\dfrac{\nu_2}{2} + k, \dfrac{\nu_1}{2} - k\right)}{B\left(\dfrac{\nu_2}{2}, \dfrac{\nu_1}{2}\right)I_w\left(\dfrac{\nu_2}{2}, \dfrac{\nu_1}{2}\right)}\left(\frac{SSW}{SSB}\right)^k$$

where $\nu_2 = A - 1$ (the between-group degrees of freedom) and SSB is the between-group sum of squares $n\Sigma(\bar{y}_a - \bar{y}.)^2$. The corresponding value of w is $SSB/(SSW + SSB)$. The logical replacement for the two quantities ν_2 and SSB when we shrink toward a regression line rather than a grand mean is $\nu_2 = A - 2$ and $nSSE = n\Sigma(\bar{y}_a - \hat{y}_a)^2$. Thus, for the case $\nu_0 = 0$, $r = 1$, our formula (A.3) and the Box–Tiao formula (7.2.27) correspond.

Finally, the formulas for $E(\lambda^{-1}|y)$ and $E(\Psi\lambda^{-1}|y)$ are obtained by first multiplying $p(\lambda, \Psi|y)$ by λ^{-1}, then integrating out λ, multiplying the result by either 1 or Ψ, and finally integrating out Ψ. The integrations are very similar to those performed in the derivation of (A.3). The results are

$$
E(\lambda^{-1}|y) = \frac{(\nu_0 m_0 + SSW)B\left(\frac{\nu_2}{2} + 1 - r, \frac{\nu_0 + \nu_1}{2} + r - 2\right)}{(\nu_0 + \nu_1 + \nu_2 - 2)B\left(\frac{\nu_2}{2} + 1 - r, \frac{\nu_0 + \nu_1}{2} + r - 1\right)}
$$

$$
\times \frac{I_w\left(\frac{\nu_2}{2} + 1 - r, \frac{\nu_0 + \nu_1}{2} + r - 1\right)}{I_w\left(\frac{\nu_2}{2} + 1 - r, \frac{\nu_0 + \nu_1}{2} + r - 1\right)} \tag{A.4}
$$

and

$$
E(\Psi\lambda^{-1}|y) = \frac{(\nu_0 m_0 + SSW)^2}{(SSE)(\nu_0 + \nu_1 + \nu_2 - 2)} \frac{B\left(\frac{\nu_2}{2} + 2 - r, \frac{\nu_0 + \nu_1}{2} + r - 3\right)}{B\left(\frac{\nu_2}{2} + 1 - r, \frac{\nu_0 + \nu_1}{2} + r - 1\right)}
$$

$$
\times \frac{I_w\left(\frac{\nu_2}{2} + 2 - r, \frac{\nu_0 + \nu_1}{2} + r - 3\right)}{I_w\left(\frac{\nu_2}{2} + 1 - r, \frac{\nu_0 + \nu_1}{2} + r - 1\right)} \tag{A.5}
$$

REFERENCES

Box, G. E. P., and Tiao, G. C. (1968), "Bayesian Estimation of Means for the Random-Effect Model," *Journal of the American Statistical Association*, **63**, 174–181.

Box, G. E. P., and Tiao, G. C. (1973), *Bayesian Inference in Statistical Analysis*, Addision-Wesley, Reading, MA.

Casella, G. (1985), "An Introduction to Empirical Bayes Data Analysis," *American Statistician*, **39**(2), 83–87.

Dempster, A. P., Laird, N. M., and Rubin, D. B. (1977), "Maximum Likelihood from Incomplete Data via the EM Algorithm" (with discussion), *Journal of the Royal Statistical Society B*, **39**, 1–38.

Dempster, A. P., Rubin, D. B., and Tsutakawa, R. K. (1981), "Estimation in Covariance Components Models," *Journal of the American Statistical Association*, **76**, 341–353.

Fay, R. E., III, and Herriot, R. A. (1979), "Estimates of Income for Small Places: An Application of James-Stein Procedures to Census Data," *Journal of the American Statistical Association*, **74**, 269–277.

Lindley, D. V., and Smith, A. F. M. (1972), "Bayes Estimates for the Linear Model" (with discussion), *Journal of the Royal Statistical Society B*, **34**, 1–42.

Morris, C. N. (1983), "Parametric Empirical Bayes Inference: Theory and Applications" (with discussion), *Journal of the American Statistical Association*, **78**, 47–65.

Nebebe, F. (1984), "Bayes and Empirical Bayes Shrinkage Estimates for Regression Coefficients, with Application to WISC Data," Ph.D. dissertation, Queen's University.

Ngai, J. (1982), "Bayesian Analysis of the Unbalanced One-Way ANOVA Model and Comparison with Empirical Bayes Approach," M.Sc. thesis, Queen's University.

Strawderman, W. E. (1971), "Proper Bayes Minimax Estimators of the Multivariate Normal Mean," *Annuals of Mathematical Statistics*, **42**, 385–388.

Stroud, T.W.F. (1984), "Bayesian Shrinkage Estimates for Regression Coefficients in m Populations," *Communications in Statistics.: Theory and Methods*, **13**, 2085–2109.

PART 4

Organizational Experiences with Small Area Techniques

Using Model-Based Estimation to Improve the Estimate of Unemployment on a Regional Level in the Swedish Labor Force Survey

C.-M. Cassel, K.-E. Kristiansson, G. Råbäck, and S. Wahlström
Statistics Sweden

ABSTRACT

This paper is one result of a project designed to apply model-based estimation on a local level. We compare the three estimators (1) the sample mean, $\hat{P}_{1q}$, (2) the uncorrected synthetical estimator, $\hat{P}_{2q}$, (3) the corrected synthetical estimator, $\hat{P}_{3q}$, with respect to estimated variances. The object is to estimate, at the municipal level, the proportion of unemployed according to the definition used in the Swedish Labor Force Survey (AKU). In doing so we make use of auxiliary information such as sex, age, and whether or not the individual is registered at the National Labor Market Board (AMS). This information is used to develop various models regarding the relationship with unemployment as measured by the AKU. The estimators $\hat{P}_{2q}$ and $\hat{P}_{3q}$ are to a certain extent based on these models. The estimator $\hat{P}_{2q}$ has the smallest variance but also an unknown bias for domains of study outside the coverage of the model. The estimator $\hat{P}_{3q}$ has a smaller variance than $\hat{P}_{1q}$ and is unbiased even where the model is not applicable. The comparisons show that for the given sample size, the only estimator that has a variance small enough for practical use is $\hat{P}_{2q}$, the uncorrected synthetic estimator. (However, as has already been pointed out, $\hat{P}_{2q}$ is biased and therefore its practical usefulness is doubtful.) By means of specially designed evaluation studies it may be possible to get information on the model bias.

141

1. BACKGROUND

The need for reliable information for regional and local community planning often makes it desirable to conduct very large surveys or censuses of entire populations. This is usually too expensive, however, and accordingly one must seek alternative solutions. Internationally, especially in the United States and Canada, there is a great interest in developing new methods for estimation in small domains of study.

1.1. Synthetic Estimation

One technique that has existed for a long time but has recently been further developed and improved is so-called synthetic estimation. The word "synthetic" is used with several different meanings, but the description given in Gonzalez (1973) has become generally accepted: "An unbiased estimate is obtained from a sample survey for a large area; when this estimate is used to derive estimates for subareas on the assumption that the small areas have the same characteristics as the larger area, we identify these estimates as synthetic estimates." According to Särndal (1984), this definition expresses the two ideas that form the basis for the technique known as synthetic estimation: that the estimate is a mixture of subestimates and that these subestimates are not genuine but rather based on a model.

The following characteristics make the technique suitable for estimation in regional domains of study:

(a) For a certain sample size in a domain one gets an estimator with better precision than the corresponding estimator obtained with the traditional technique in which only information from the sample appearing in the actual domain is used.

(b) It is possible to get estimates even for domains in which the number of observations in the sample is so small that with traditional technique one cannot calculate an estimate at all.

In synthetic estimation the following requirements must be fulfilled:

(c) For each unit in the population under study there must exist auxiliary information correlated to the variable being studied. The higher the correlation, the better the estimator.

(d) The "model assumption" must hold: the relations that can be observed in large population groups must also hold for small domains.

The population under study is split into a number of groups with respect to the auxiliary variable. Information about the relation between the auxiliary variable and the variable being studied is obtained from the sample.

A synthetic estimate for each domain of study is then calculated by weighting and summing the estimates from each subgroup of the domain. The weight consists of the number of units in the population subgroup in the domain. The procedure is based on the "model assumption" mentioned previously: that the relations which hold for the large groups of the population also hold for the small domains of study. If this model assumption is true, synthetic estimation has its strength in "borrowing" information from large groups for use in small domains in which the basis for estimation is in itself insufficient. On the other hand, if the model assumption is not true, the technique of synthetic estimation in its "pure" form gives biased estimates. The estimators described in Särndal (1984), however, are unbiased even if the model assumption is false; thus, he has improved the original technique considerably. The unbiasedness property is obtained by adding a term that might be of low precision in small domains of study where the sample size is small.

1.2. Application

The study described in this paper is an attempt to improve the estimate of unemployment according to the definition used in AKU (the Swedish Labor Force Survey) in local domains of study. AKU is a sample survey in which a monthly sample of 22,000 individuals are interviewed about their status in the labor market. It is difficult to estimate the ratio of unemployment in a small area, because the number of unemployed in the sample in that area is small, indeed often zero. The total sample size might be considerable, but even so, the estimates for relatively large domains of study have low precision. Generally, at the municipal level it is useless to try to make estimates by means of the technique presently in use, which involves the ordinary sample mean (see below for a definition).

The auxiliary information used in this study consists of statistics on the number of applications for work registered by the employment exchange in each municipality (AMS). A description of relations and differences between the statistics of AMS and AKU is given in Section 3.3. The calculations in the present study refer to three different months, for which the AKU-samples have been matched against the corresponding AMS-registers. A more detailed description of the empirical material can be found in Section 3.2.

The study was preceded by a Monte Carlo simulation study on the same empirical material, see Cassel, Råbäck, and Särndal (1983). Estimates were calculated for clusters of municipalities (so-called "counties"). The results showed that it should be possible to construct an unbiased synthetic estimator about 25% more efficient than the one now used, measured in terms of standard deviations.

2. DESCRIPTION OF ESTIMATORS

The following paragraphs give the notation used to describe the AKU and AMS status of an individual.

The property of being unemployed according to AKU is denoted AKU1; that of being employed is denoted AKU0. The property of being registered at AMS is denoted AMS1; that of being unregistered is denoted AMS0.

The number of AMS1 in regional domains is known from the monthly statistics from AMS. The number of AMS0 is calculated by subtraction from the population figures. Our aim is to estimate the proportion of unemployed in the population aged 16–64 in counties and municipalities according to the AKU definition. Let P_0 (AKU1) denote that proportion.

We have studied three types of estimators, which in principle have the following characteristics:

$\hat{P}_{1q}$, *the sample mean*, in each domain q without auxiliary information, that is,

$$\hat{P}_{1q} = \frac{1}{n_q} \sum_{k \in s_q} Y_k$$

where n_q = the number of units in the sample that fall into domain q
s_q = the set of units in the sample s which fall into q
$Y_k = \begin{cases} 1 & \text{if individual } k \text{ is AKU1} \\ 0 & \text{otherwise} \end{cases}$

The estimator is unbiased and uses only the units belonging to s_q. The precision decreases when n_q becomes small.

$\hat{P}_{2q}$, *the "uncorrected" synthetic estimator*. This is the original synthetic estimator. It uses AMS-register figures. The estimate for a certain domain is based on relations observed for a larger group and assumed to hold true for that domain. The estimator is biased if the assumption is false. The estimator makes use of observations from the entire sample.

The population is split into H subgroups using the auxiliary information. In each subgroup h the proportion of unemployed according to the AKU definition among the units belonging to subgroup h is estimated. This parameter will be denoted θ_h, $h = 1, \ldots, H$.

The estimator $\hat{P}_{2q}$ is formed as a weighted sum of these estimated parameters $\hat{\theta}_h$. The weights are the respective proportions of the domain belonging to a subgroup; that is,

$$\hat{P}_{2q} = \sum_{h=1}^{H} (N_{hq}/N_{.q})\hat{\theta}_h$$

with

$$\hat{\theta}_h = \sum_{k \in s_h} Y_k / n_h$$

where N_{hq} = the number of individuals in the population belonging to subgroup h and domain q

N_q = the number of individuals in the population belonging to domain q

n_h = the number of individuals in the sample belonging to subgroup h

s_h = the part of the sample s falling into subgroup h

The number of subgroups, H, and their definitions depend on the model assumptions. In section 3 we describe the different models we have tried.

$\hat{P}_{3q}$, the *"corrected" synthetic estimator*, also uses the auxiliary variable. It is based on a model assumption, like $\hat{P}_{2q}$, but it also estimates the deviations from the model in each domain. Since it takes the residuals into account, the estimator is unbiased (asymptotically) even if the model is defective. The estimator consists of two parts, a pure model part, which is identical to $\hat{P}_{2q}$, and a correction part. In the model part, information from the entire sample is used, while in the correction part information from the domain and the entire sample is used. We will compare $\hat{P}_{2q}$ and $\hat{P}_{3q}$ with $\hat{P}_{1q}$. We have

$$\hat{P}_{3q} = \hat{P}_{2q} + \sum_{s_q} e_k / \Pi_k$$

where Π_k is the inclusion probability of unit k, and $e_k = Y_k - \hat{Y}_k$. $\hat{Y}_k = \hat{\theta}_h$ if unit k belongs to group h, $h = 1, \ldots, H$.

3. MODELS

In this section we discuss in detail the relationship between AMS and
AKU in order to use this relationship to build a number of models; we
then examine these models.

3.1. Covariation between AMS and AKU

Figure 1 shows that the AMS variable seems to be strongly correlated
over time with the AKU variable.

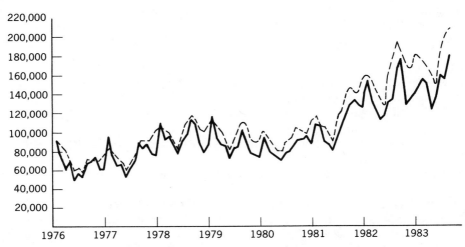

Figure 1. The number of unemployed measured by AMS (dotted line) and AKU (solid
line) in 1976–1983.

3.2. Empirical Support for the Study

In order to study in some detail the relation between the number of
unemployed according to AKU and the applications for work registered
at AMS, data from a matching of AKU and AMS were used. To the
AKU samples from November 1982, January 1983, and March 1983 were
added data from the AMS register of applicants for work. The AKU data
were matched against the AMS register as it appeared on the Tuesday
following the week selected for measurement in the AKU survey. However, the version of the register from which the monthly AMS statistics
are computed refers to the last working day of the month. Out of the

62,661 individuals in the AKU samples, 2128 were identified as applicants for work in the AMS register. Because of nonresponse in AKU, the number of individuals both interviewed in the AKU survey and registered at the AMS was reduced to 1926.

The number of registered individuals in the population broken down by sex, age, and municipality were obtained from the 1982/83 RTP (the Register of the Total Population) and the total counts from the AMS registers.

3.3. The Relation between AMS and AKU

The information on the relation between AMS and AKU is to a large extent based on data from the matching. Before discussing the results of this process, we first give a short description of the two sources of statistics.

As was previously mentioned, AKU is a sample survey covering 22,000 individuals each month. By means of telephone interviews, data are collected about the emloyment status of the individual during a certain week. An individual is considered unemployed if she/he was unemployed during the measurement week and (1) had applied for work or was waiting for the outcome of job applications made during the last 60 days, or (2) was waiting for reemployment on a job from which she/he was temporarily laid off without salary, or (3) would have looked for work if she/he had not been ill. The nonresponse rate in the AKU survey is about 6%.

The AMS statistics are based on a total count of the number of individuals without work who are registered as job applicants and are ready to start work within 10 days. The basic element of the register is a "statistics card" that is drawn up for each individual by the employment exchange.

Each time a person revisits the employment exchange, the register is updated. An applicant is given an appointment for a new visit in approximately four weeks. Thus, about four weeks will pass by until the employment exchange learns whether an applicant has become employed, if such a change of status has taken place. Because of this time lag, the information in the register can be out of data for some units.

In these aspects the two sources of statistics on unemployment differ. Table 1 summarizes the relation between AMS and AKU. The figures are based on data from the matching procedure. Thus they can be seen as reliable estimates since approximately 66,000 units were involved in the matching. Among those registered by AMS, 62% are unemployed according to AKU.

Table 1.   The Relation between AMS and AKU[a]

	AMS		
AKU	AMSO (Not Registered Applicants for Work)	AMS1 (Registered Applicants for Work)	Total
AKUO (Employed)	96.2	1.1	97.3
AKU1 (Unemployed)	0.9	1.9	2.7
Total	97.0	3.0	100.0

[a] This table shows percentage figures for the cross-classification of AMS status with AKU status.

The first model (later called model A) used to construct a synthetic estimator is based on $H = 2$ groups. More precisely, the parameters $\theta_1 - \theta_1$ is the proportion of AMS1 who are also AKU1—and $\theta_2 - \theta_2$ is the proportion of AMSO who are also AKU1—estimated from the matched data to be 62% and 0.9% on a national level should be about the same for each municipality. In municipalities where these proportions do not coincide with the national values, the synthetic estimator will yield biased estimates.

In order to judge how applicable the model is, it seems necessary to explore the causes of the discrepancies between AMS and AKU. If there were no difference between the AMS and AKU definitions of unemployment, the off-diagonal cells in Table 1 would be zero. However, the off-diagonal cells in Table 1 are nonzero. In the next section we seek explanations for this phenomenon. These considerations then form a basis for model building.

3.4.   Some Causes of Differences between the AMS and AKU Statistics and the Impact of These Differences on the Parameters θ_1 and θ_2

We will discuss two possible causes of the observed differences between the AMS and AKU statistics.

(a)   Differences in Definitions between AMS and AKU, and Routines for Updating the AMS Register
Due to the time lag in the updating of the register of applicants, the register contains a number of individuals who are no longer looking for

work. The number of these individuals depends on the conditions on the labor market. From the AMS register we know that in 1982 the number of individuals in the entire Swedish population who ceased to look for work during a four-week period varied between 20,000 and 40,000. This variation follows a seasonal pattern. The consequences ought to be that θ_1 increases when the number of individuals no longer looking for work decreases. However, we judge that θ_1 does not vary more than $\pm 1\%$ during the year as a result of this seasonal pattern. If there was no time lag at all, that is, if all registered individuals were really still looking for work, θ_1 would probably rise to 65% at most.

Among the registered applicants in the matched data set, 62% were estimated to be unemployed according to AKU, while 26% were employed and 12% were classified as not in the labor force. From the previous discussion we see that the routines for updating the register can explain this discrepancy only to a minor extent. The definitions of unemployment are such that it is quite possible for a person to be looking for work in the sense of AMS and still be employed in the sense of AKU. By "applicants without work" AMS means persons without work and ready to take a job within 10 days. To this group are also assigned people whose employment ceases within 10 days and people having a steady sideline. On the other hand, AKU considers people employed if these have worked at least 1 hr during the week of measurement. Applicants who are referred to short-term jobs (less than 10 days) are considered by AMS as applicants without work, despite the fact that they may have been working full-time. Thus it is possible that all individuals in this group might be applicants with short-term work. Since we do not know the number of applicants with temporary jobs, we cannot estimate to what extent their existence would explain the 26% employed. Persons who are new on the labor market or are rejoining it probably constitute a large share of the applicants with temporary jobs.

(b) Causes due to Age-Specific Reasons
Our empirical data show that the proportion of registered applicants who are employed is larger among individuals who are less than 55 years old and among women than among older persons and males (Table 2). As for the registered applicants who are unemployed according to AKU, we believe that about 40% can be explained as consisting of people visiting the exchange just to indicate their existence. When firms must reduce their personnel, they frequently offer early retirement pensions preferentially to people who have reached 58 years of age. Those who accept this offer will be registered at the labor exchange and will show up there until their 60th birthday when they can obtain their national early retirement

Table 2. The Proportions of People Unemployed, Employed, and Not in the Labor Force, among Individuals Registered by AMS, Classified by Sex and Age (%)

	Sex					
	Male			Female		
Age	Unemployed θ_1	Employed	Not in the Labor Force	Unemployed θ_1	Employed	Not in the Labor Force
16–24	66.8	25.4	7.8	57.5	33.6	8.9
25–54	67.3	26.3	6.4	57.3	31.0	11.7
55–64	68.2	11.7	20.1	55.4	14.0	30.6

pensions. In principle such persons are at the disposal of the labor market, despite the fact that they are registered only to secure their living. But one can expect these people, when interviewed, to declare that they have left the labor market. This can be seen from Table 2.

As can be seen from Table 1, many unemployed individuals are not registered by AMS. Of course it is possible to be looking for work without being registered: according to AKU (mean value of 1983) about 10% of the unemployed (15,000 individuals) had been looking for work without being registered by AMS. Even among those who, when interviewed in AKU, declared that they were in the AMS register, there were probably some who were looking for work without really being in that register.

A reasonable guess is that it is, above all, newcomers to the labor market who are looking for work without being registered by AMS, since there is then no financial inducement. Table 3 shows the proportion unemployed among people who are not registered as applicants (θ_2).

As can be seen from Table 3, θ_2 seems to vary with age. Since θ_1 and θ_2 vary between sex and age groups, the model assumption will probably be

Table 3. Proportion of Unemployed Among Individuals Not Registered as Applicants by AMS (%)

Age	Male	Female
16–24	1.5	1.7
25–54	0.8	0.7
55–64	0.5	0.4

more realistic if the population were divided into sex and age groups as well (see model E below). However, our results show that the estimated variance does not necessarily decrease when the number of groups, H, increases.

3.5. An Alternative Classification of the Population

An alternative classification of the population that puts more emphasis on the labor market situation of the individual can be constructed from the kind of financial compensation involved. For example, the applicants can be divided into three groups. Applicants in the first group get labor market compensation in cash (KAS), those in the second group get another type of compensation called fund compensation, and those in the third group get no compensation. The proportion of unemployed according to AKU among individuals registered by AMS in these groups are 60%, 66%, and 54%, respectively. The overall proportion is 62%.

The different kinds of compensation provide at the same time a classification of the individual's association to the labor market. Applicants getting fund compensation have the strongest association to the market. Applicants without compensation are to a large extent newcomers to the market, while applicants with KAS could be classified somewhere in between.

The distribution of the three groups probably differs in different local labor markets. Compare, for example, a municipality in which the dominant industry has been closed down and a municipality in which the major part of the applicants are students. The distribution of the groups probably also varies by season: one might expect a large number of newcomers to the market right after the summer. The groups also vary by the state of the market: in a depression even the well established are affected, and this means that the applicants then include a larger number of individuals entitled to compensation.

This classification of the population aims at achieving a more realistic model but also at diminishing the variation of the relations over time, thereby increasing the possibility of estimating the parameters with high precision (see model E).

3.6. Description of the Models Used

The different models are distinguished by the number of groups H into which the population is divided and the way in which the auxiliary variables are used.

Model A divides the population into two groups: that is, $H = 2$. One group consists of AMSO, the other of AMS1.

Model B uses $H = 48$ groups. The groups are formed by subdividing AMS0 and AMS1 by county. (Sweden is divided into 24 counties.) The model assumes that the relation between the status of AMS and AKU is different in each county.

Model C uses information on AMS status, sex, and age, $H = 12$. There are six age/sex groups crossed by two AMS status categories.

We also tried two models C^1 and C^{11}. Model C^1 was a more elaborate version of C; county information was added. Model C^{11} was less elaborate than C, using only information on sex and age. We found that model C was better in the sense of giving on the average smaller estimated confidence intervals than C^{11} and that no substantial improvement was provided by model C^1. Thus, we will only report the behavior of the estimators under model C.

Model D uses AMS status and information on the so-called H-regions (Sweden is divided into 8 H-regions, where each region consists of municipalities with a similar structure), $H = 16$.

Model E has $H = 4$. Group 1 consists of AMS0. The remaining three groups are formed by further division of the group AMS1 into individuals entitled to fund compensation, individuals entitled to compensation in cash, and individuals not entitled to any compensation at all.

4. RESULTS

4.1. Estimates of the Model Parameter

The synthetic estimators $\hat{P}_{2q}$ and $\hat{P}_{3q}$ are built as linear combinations of estimates of the proportion of unemployed according to AKU within specified groups. Let θ_h denote the proportion of AKU1 in group h with the corresponding estimate $\hat{\theta}_h$. We will start by giving these estimates for most of the models studied. If the auxiliary information is effective, a large difference should be seen between the θ_hs in the different groups. In this application the θ_hs should also be stable for different points of time. Table 4 shows how $\hat{\theta}_h$ varies.

In model A, the estimates $\hat{\theta}_1$ and $\hat{\theta}_2$ relate to the proportion of AKU1 among AMS1 and AKU1 among AMS0, respectively.

In model B, the estimates $\hat{\theta}_{1j}$ and $\hat{\theta}_{2j}$ relate to the same proportions as in model A but for county j, $j = 1, \ldots, 24$.

In model C, the estimate $\hat{\theta}_{21}$ relates to the proportion of AKU1 among AMS0 in the group of men aged 16–24 years, $\hat{\theta}_{22}$ to the corresponding

Table 4. Estimated θ_h in Different Groups According to Models A, B, C, and E

Model A

	θ_1	θ_2
November	0.607	0.007
January	0.595	0.008
March	0.657	0.007

Model B

j	November $\hat{\theta}_{1j}$	November $\bar{\theta}_{2j}$	January $\hat{\theta}_{1j}$	January $\bar{\theta}_{2j}$	March $\hat{\theta}_{1j}$	March $\hat{\theta}_{2j}$
1	0.525	0.004	0.490	0.010	0.584	0.006
2	0.655	0.005	0.489	0.009	0.775	0.004
3	0.728	0.009	0.719	0.014	0.487	0.014
4	0.594	0.005	0.376	0.006	0.657	0.006
5	0.709	0.004	0.596	0.006	0.582	0.004
6	0.575	0.006	0.654	0.019	0.754	0.005
7	0.752	0.004	0.631	0.006	0.710	0.007
8	0.623	0.008	0.624	0.007	0.753	0.005
9	0.716	0.001	0.532	0.003	0.722	0.001
10	0.740	0.012	0.636	0.003	0.716	0.009
11	0.605	0.010	0.695	0.008	0.580	0.008
12	0.490	0.004	0.716	0.008	0.495	0.009
13	0.585	0.006	0.607	0.007	0.570	0.010
14	0.483	0.015	0.562	0.006	0.743	0.011
15	0.730	0.011	0.416	0.004	0.686	0.006
16	0.604	0.003	0.512	0.011	0.668	0.006
17	0.623	0.004	0.612	0.007	0.697	0.007
18	0.700	0.010	0.693	0.009	0.588	0.012
19	0.403	0.005	0.523	0.013	0.709	0.005
20	0.706	0.005	0.461	0.007	0.716	0.001
21	0.662	0.007	0.810	0.007	0.630	0.009
22	0.487	0.003	0.660	0.013	0.709	0.001
23	0.425	0.008	0.713	0.011	0.662	0.011
24	0.603	0.010	0.540	0.006	0.813	0.013

Model C

	$\hat{\theta}_{11}$	$\hat{\theta}_{12}$	$\hat{\theta}_{13}$	$\hat{\theta}_{14}$	$\hat{\theta}_{15}$	$\hat{\theta}_{16}$
November	0.665	0.657	0.709	0.576	0.579	0.432
January	0.604	0.682	0.620	0.589	0.519	0.547
March	0.735	0.684	0.716	0.560	0.620	0.648

	$\hat{\theta}_{21}$	$\hat{\theta}_{22}$	$\hat{\theta}_{23}$	$\hat{\theta}_{24}$	$\hat{\theta}_{25}$	$\hat{\theta}_{26}$
November	0.016	0.005	0.003	0.016	0.007	0.002
January	0.014	0.010	0.002	0.017	0.009	0.002
March	0.014	0.010	0.003	0.018	0.006	0.002

Table 4. (*Continued*)

Model E	$\hat{\theta}_0$	$\hat{\theta}_{11}$	$\hat{\theta}_{12}$	$\hat{\theta}_{13}$
November	0.007	0.571	0.637	0.572
January	0.008	0.618	0.635	0.498
March	0.007	0.616	0.717	0.552

proportion among men aged 25–54 years, $\hat{\theta}_{23}$ to the corresponding proportion among men aged 55–64 years. The estimates $\hat{\theta}_{24}$–$\hat{\theta}_{26}$ relate to the corresponding proportions among women in the same age groups as the men. The estimates $\hat{\theta}_{11}$–$\hat{\theta}_{16}$ relate to the proportion of AKU1 among AMS1 in the same age/sex groups as before.

In model E, $\hat{\theta}_2$ relates to the proportion of AKU1 among AMS0, $\hat{\theta}_{11}$ to the proportion of AKU1 among AMS1 with KAS $\hat{\theta}_{12}$ to the corresponding proportion but with fund compensation, and $\hat{\theta}_{13}$ to the corresponding proportion without compensation.

It should be pointed out that the previous numbers are estimates subject to sampling variations and not true values of a parameter. The precision decreases when the number of groups increases because the estimate is then based on fewer observations. This reveals a difficult problem: the model must conform with reality (and thus be fairly detailed), but from considerations regarding the precision it seems at the same time desirable to have a small number of parameters. It is evident from Table 5 that the most efficient auxiliary information is that already introduced in model A. How far the additional auxiliary information in the other models will influence the accuracy is an important question. The introduction of additional information in order to reduce the model bias may lead to an unacceptable loss in precision.

4.2. Estimates of Unemployment at the Municipal Level

The main object of our study was to estimate $P_0(AKU1)$ using municipalities as domains of study. As was previously mentioned, the AKU survey is based on a monthly sample of 22,000 individuals drawn from the population of individuals aged 16–64 years living in Sweden. It is stratified according to county, sex, marital status, and citizenship. Within a stratum, the individuals are selected by simple random sampling. The standard errors are calculated using a straightforward elaboration of the well-known formula for the variance of a stratified sample. We decided not to burden our paper with the formula. Table 5 contains

Table 5. Estimates of P_0(AKU1) with Estimated Standard Errors (Within Brackets) for Selected Municipalities using Models A–D for AKU in November 1982

Municipality	Number of Inhabitants 16–64 Years	n_o	$\hat{P}_{1q}$	A		B		C		D	
				$\hat{P}_{2q}$	$\hat{P}_{3q}$	$\hat{P}_{2q}$	$\hat{P}_{3q}$	$\hat{P}_{2q}$	$\hat{P}_{3q}$	$\hat{P}_{2q}$	$\hat{P}_{3q}$
Botkyrka	44,199	175	2.24 (1.12)	1.78(0.07)	2.28(0.88)	1.37(0.17)	2.37(0.97)	1.92(0.08)	2.23(0.87)	1.39(0.17)	2.36(0.97)
Stockholm	422,694	1,358	0.79 (0.24)	1.63(0.06)	0.86(0.23)	1.24(0.16)	0.90(0.23)	1.68(0.07)	0.93(0.24)	1.25(0.16)	0.90(0.23)
Tierp	12,028	57	0.00	2.05(0.07)	−0.07(0.98)	2.02(0.44)	−0.02(1.05)	2.11(0.07)	−0.21(1.14)	2.16(0.15)	−0.06(1.06)
Katrineholm	19,715	108	3.42 (1.75)	2.71(0.09)	5.20(1.33)	3.33(0.51)	5.20(1.28)	2.77(0.09)	5.21(1.32)	2.88(0.18)	5.24(1.31)
Kinna	5,897	21	3.99 (4.27)	2.33(0.08)	0.09(3.19)	2.17(0.41)	0.24(4.36)	2.36(0.08)	0.16(3.93)	2.27(0.25)	0.03(3.20)
Gnosjö	5,514	13	7.73 (7.39)	1.42(0.06)	2.88(3.67)	1.25(0.27)	2.41(2.72)	1.50(0.07)	3.11(3.94)	1.48(0.13)	2.72(3.22)
Gotland	34,492	636	1.95 (0.54)	1.94(0.07)	2.19(0.47)	2.16(0.46)	2.16(0.47)	2.03(0.08)	2.27(0.47)	1.69(0.21)	2.18(0.47)
Malmö	149,195	450	3.65 (0.88)	3.81(0.12)	4.32(0.72)	4.11(0.43)	4.29(0.77)	3.85(0.12)	4.34(0.77)	3.81(0.36)	4.27(0.77)
Göteborg	278,468	910	1.88 (0.45)	2.81(0.09)	2.56(0.39)	2.67(0.33)	2.55(0.39)	2.88(0.09)	2.70(0.38)	2.90(0.28)	2.47(0.39)
Laxå	5,218	15	6.51 (6.36)	3.38(0.10)	4.78(2.96)	3.22(0.55)	4.71(2.84)	3.42(0.11)	4.93(3.18)	3.59(0.22)	4.77(2.60)
Arjeplog	2,494	8	19.96(14.11)	3.36(0.12)	11.88(5.54)	4.18(0.56)	11.87(5.51)	3.85(0.13)	11.58(5.37)	3.83(0.35)	11.73(5.41)

the estimates, and estimates of the corresponding standard deviation for some selected municipalities in November 1982.

For each municipality, Table 5 shows the number of inhabitants, the number of individuals in the sample, and the estimate $\hat{P}_{1q}$ followed by the estimated standard deviation of $\hat{P}_{1q}$ with brackets. Then follow estimates and standard deviations of $\hat{P}_{2q}$ and the corresponding quantities for $\hat{P}_{3q}$. The synthetic estimates $\hat{P}_{2q}$ and $\hat{P}_{3q}$ have been calculated for four different models, namely, models A–D. For each model we find that the standard error of $\hat{P}_{2q}$ is considerably smaller than that of $\hat{P}_{1q}$ or $\hat{P}_{3q}$. The standard error of $\hat{P}_{2q}$ increases with the number of groups. $\hat{P}_{3q}$ has a smaller standard deviation than $\hat{P}_{1q}$ in the models that use AMS information.

For each municipality it is possible to calculate an interval for $P_0(\text{AKU1})$. For example, for Botkyrka we get for $\hat{P}_{1q}$ the interval $(0.05-4.34)\%$, for $\hat{P}_{2q}$ model A, the interval $(1.65-1.91)\%$ and for $\hat{P}_{3q}$ model A, the interval $(0.56-4.00)\%$ using standard 95% confidence limits. The interval for $\hat{P}_{3q}$ is somewhat shorter than that for $\hat{P}_{1q}$. However, only the interval for $\hat{P}_{2q}$ is sufficiently short: the other two are much too wide. Comparisons for the other models give similar results. It must be kept in mind that the estimator $\hat{P}_{2q}$ is heavily dependent on the model and can be biased if the model does not hold. In that case the confidence level of the interval is unknown due to the unknown bias.

Since the intervals for $\hat{P}_{1q}$ and $\hat{P}_{3q}$ are so wide, the lower endpoint of the interval is often found below zero. The estimate $\hat{P}_{3q}$ might even turn out negative (cf., the municipality of Tierp). Such unacceptable estimates occur when the sample from the domain contains many units with the property AMS1 and AKU0. The influence of the correction term in $\hat{P}_{3q}$ then becomes unreasonably large. The corresponding $\hat{P}_{1q}$ estimate often takes the value 0 and thus it is impossible to form a confidence interval for $\hat{P}_{1q}$.

Model A gives confidence intervals that are slightly shorter than for model C. Models A and C give shorter intervals than models B and D. If we measure the goodness of the models by their ability to produce short intervals, then model A is the best. However, model C, which uses information on AMS status, sex, and age is more detailed than model A, and one would therefore expect model C to be more realistic than model A. Instead a certain amount of precision is lost, because of the increase in the number of parameters to be estimated.

Table 6 shows the estimates of $\hat{P}_{1q}$, $\hat{P}_{2q}$, and $\hat{P}_{3q}$ for the three measurement periods for a number of municipalities. In the first column we can see how $P(\text{AMS1})$ varies between the months. For completeness

Table 6. Comparisons of P(AMS1) and of the Estimates of P(AKU1) for Selected Municipalities[a]

		Models							
		A		B		C		D	
P(AMS1)	$\hat{P}_{1q}$	$\hat{P}_{2q}$	$\hat{P}_{3q}$	$\hat{P}_{2q}$	$\hat{P}_{3q}$	$\hat{P}_{2q}$	$\hat{P}_{3q}$	$\hat{P}_{2q}$	$\hat{P}_{3q}$
Sollentuna (115)									
N	0.90	2.27	1.20	2.95	0.86	2.99	1.79	3.01	0.87	2.98
J	1.16	0.83	1.51	1.21	1.52	1.10	1.65	1.32	1.57	1.11
M	1.01	3.74	1.41	3.66	1.25	3.83	1.52	3.70	1.19	3.85
Stockholm (1325)									
N	1.62	0.79	1.63	0.86	1.24	0.90	1.68	0.93	1.25	0.90
J	1.65	1.90	1.80	2.34	1.76	2.27	1.88	2.46	1.81	2.27
M	1.51	1.08	1.73	1.58	1.53	1.54	1.79	1.69	1.47	1.55
Enköping (100)									
N	2.94	1.29	2.43	2.58	2.43	2.72	2.48	2.62	2.57	2.67
J	3.34	1.85	2.79	1.40	2.51	1.50	2.89	1.45	2.87	1.36
M	3.00	3.25	2.70	1.97	2.71	1.69	2.79	2.08	2.85	1.83
Nyköping (179)									
N	2.47	1.89	2.14	2.58	2.69	2.67	2.20	2.67	2.26	2.62
J	2.88	2.94	2.52	2.67	3.45	3.60	2.61	3.86	2.59	3.70
M	2.44	4.42	2.34	3.84	2.51	4.12	2.42	3.84	2.47	3.75
Hultsfred (51)									
N	4.50	8.22	3.36	5.32	3.76	4.16	3.43	4.96	3.58	4.91
J	4.49	2.29	3.46	3.59	3.37	3.68	3.54	3.69	3.60	3.69
M	4.18	6.37	3.47	4.90	3.65	4.73	3.55	4.90	3.67	4.76
Gotland (636)									
N	2.13	1.95	1.94	2.19	2.16	2.16	2.03	2.27	1.69	2.18
J	2.25	1.92	2.15	2.08	2.10	2.11	2.27	2.20	1.90	2.14
M	4.18	6.37	3.42	4.90	3.65	4.73	3.55	4.90	3.67	4.76
Malmö (450)									
N	5.24	3.64	3.81	4.22	4.11	4.29	3.85	4.39	3.81	4.26
J	5.49	3.84	4.05	4.64	4.63	4.77	4.15	4.77	4.09	4.69
M	5.40	3.87	4.26	4.41	3.92	4.40	4.32	4.65	4.11	4.38

[a] N ≡ November; J ≡ January; M ≡ March.

the estimates $\hat{P}_{2q}$ and $\hat{P}_{3q}$ have been included for each model. The number in brackets next to the name of the municipality is the average number of observations in the domain. For January and March we reach the same conclusion as for November. The large sample variations for $\hat{P}_{1q}$ and $\hat{P}_{3q}$ also create large differences in the estimates for the different months. The large variation of P(AMS1) between the months is somewhat puzzling, but generally the proportion seems to increase from November to January and decrease from January to March.

Table 7 gives a different perspective of the variation over time for the estimates: it displays the frequency of small and large differences. For $\hat{P}_{1q}$ and $\hat{P}_{3q}$, the frequency of large differences between the months is considerably higher than for $\hat{P}_{2q}$.

Table 7. The Frequency of Large and Small Differences in P(AMS1) and in Estimates of P(AKU1) between Months[a]

			Models							
			A		B		C		D	
Difference	P(AMS1)	$\hat{P}_{1q}$	$\hat{P}_{2q}$	$\hat{P}_{3q}$	$\hat{P}_{2q}$	$\hat{P}_{3q}$	$\hat{P}_{2q}$	$\hat{P}_{3q}$	$\hat{P}_{2q}$	$\hat{P}_{3q}$
From November to January										
below −3.0%	2	16	1	11	1	11	1	10	0	10
−3.0−−1.5%	7	13	3	14	4	14	4	15	4	15
−1.5−−0.3%	11	7	6	13	17	18	6	15	9	15
−0.3−0.3%	25	15	36	13	29	11	35	11	38	12
0.3−1.5%	25	9	37	13	28	11	38	12	33	13
1.5−3.0%	21	12	13	15	17	12	15	15	14	14
Over 3.0%	6	23	2	21	4	23	2	21	3	21
From January to March										
below −3.0%	1	18	1	18	1	18	1	19	1	18
−3.0−−1.5%	8	7	3	12	4	4	3	12	4	9
−1.5−−0.3%	19	9	10	13	17	9	9	13	10	12
−0.3−0.3%	50	21	64	19	40	18	66	16	63	18
0.3−1.5%	12	9	16	12	32	15	16	13	15	12
1.5−3.0%	5	13	3	13	4	13	2	11	4	12
Over 3.0%	1	19	1	11	2	10	1	12	1	11

5. CONCLUSION

A comparison of the different estimators shows that although the precision of $\hat{P}_{3q}$ is better than that of $\hat{P}_{1q}$, it is not sufficient for practical use at the municipal level. This holds true for all the models studied. From this point of view the uncorrected synthetic estimator $\hat{P}_{2q}$ is the only reasonable choice. However, $\hat{P}_{2q}$ is biased if the model does not hold, and it is difficult to obtain information about the magnitude of this bias for the different models. The value of the estimate $\hat{P}_{2q}$ in social planning at the municipal level must then be evaluated in relation to existing alternatives. In other words, is it better to use the estimate of unemployment according to AKU given by the synthetic estimator based on model C, with a possible unknown model bias, or should the AMS statistics for the municipality be used? By means of specially designed evaluation studies it may be possible to get information on the model bias in different municipalities for $\hat{P}_{2q}$ when estimating $P(\text{AKU1})$.

REFERENCES

Cassel, C-M., Råbäck, G., and Särndal, C-E. (1983), "Synthetic Estimation. A Monte Carlo Study of the Possibility to Use AMS-Data as Auxiliary Information in AKU" (in Swedish), Unpublished memo, Statistics Sweden.

Gonzalez, M.E. (1973), "Use and Evaluation of Synthetic Estimates," *Proceedings of the Social Statistics Section of the American Statistical Association*, pp. 33–36.

Särndal, C-E. (1984), "Design-Consistent Versus Model-Dependent Estimation for Small Domains," *Journal of the American Statistical Association*, **79**, 624–631.

Use of Regression Techniques for Developing State and Area Employment and Unemployment Estimates

F. R. Cronkhite
U.S. Department of Labor

ABSTRACT

The Handbook method, the cornerstone of the current system for developing state and substate employment and unemployment estimates, is marked by major statistical defects, which can lead to large estimation errors. To improve the estimating system, the Bureau of Labor Statistics (BLS) undertook research to develop statistical models to replace the Handbook. This research resulted in single-equation regression models for estimating state employment and unemployment, and pooled cross-sectional time series regression models for estimating substate area employment and unemployment. The dependent variables in all models are monthly sample statistics from the Current Population Survey (CPS). The models lead to reduced estimation error for most states and areas over the test period.

1. METHODOLOGICAL OBJECTIVES

The major objectives that we have been trying to achieve in this study with statistical models are (1) to produce estimates at the state level with minimum annual errors relative to the CPS; (2) to produce estimates at

This is an abridged version of an unpublished paper which is available upon request.

160

the area level corresponding to the expected within-state cross-sectional distribution; (3) to capture the structure of the underlying economic mechanism so as to produce estimates corresponding to the expected local trend, cycle, and seasonal movements.

2. CONSTRAINTS ON STATISTICAL SYSTEM

The constraints on any statistical system that is used to derive the official BLS labor force estimates for states and areas are(1) that the CPS is the statistical standard for comparison, because it has desirable statistical properties and because it measures the official, widely held concepts and definitions of employment and unemployment; (2) that the official estimates for any state or area will be derived directly from the CPS, where the sample is large enough to produce (monthly or annual average) estimates that meet the prescribed BLS reliability standard; (3) that, following (2), CPS estimates are used directly on a monthly basis in the 11 largest states and two large metropolitan areas, and on an annual average basis in the remaining 39 states and the District of Columbia; (4) that, in the latter group of states in (3), hereafter referred to as "nondirect-use states," the monthly preliminary state estimates are annually forced, ex post, to average to the official CPS annual average estimates for the same state; and (5) that all nondirect-use substate area estimates are routinely forced to sum to the official state estimate. There is no requirement that the state estimates must sum to the national estimates. A general description of the BLS procedures that are currently used by state Employment Security Agencies to derive monthly labor force estimates for states and areas is provided in the Technical Notes of the monthly Employment and Earnings report of BLS (U.S. Department of Labor, 1985).

3. RESEARCH STRATEGY

The methodological approach we have followed has been largely dictated by the quality of local data that are available for the particular class of areas under study. The most suitable data for programmatic use for any area, of course, would be derived from a sufficiently large probability sample, since such a procedure has desirable statistical properties, like unbiasedness, acceptable variance, and susceptibility to rigorous statistical testing. The CPS is a random sampling procedure that, in addition to having such desirable statistical properties, also measures the official, widely held concepts and definitions of employment and unemployment. Consequently, we would use CPS data for all states and areas if cost was no object. Since we are dealing with the real world and limited budgets, however, we can only afford to finance the use of CPS data directly in the 11 largest states and two very large substate areas. Therefore, for all remaining nondirect-use states and areas, our research strategy has been to identify other statistical procedures that also have desirable properties, and that combine current sample data from the CPS and accounting data from the state Unemployment Insurance (UI) system, so as to maximize the use of CPS data, following constraint (1) above.*

Following this strategy, major emphasis has been focused on the classes of areas for which monthly CPS data are available. We currently have such data for all states, and some 200 Standard Metropolitan Statistical Areas (SMSAs). For both classes of areas, we have obtained very satisfactory results using single-equation regression techniques, where the dependent variables are monthly CPS estimates. Regression techniques are in keeping with our dual objective of using procedures that have desirable statistical properties and that maximize the use of CPS data.

4. REGRESSION MODEL SPECIFICATION

Each of the equations we have developed and tested is based on a multivariate linear statistical model, which is generally specified as

$$CPS = B_0 + B_1 X_1 + \cdots + B_k X_k + e \qquad (1)$$

where CPS = the dependent variable, defined as monthly estimates of employment or the unemployment rate from the CPS

$X_1 - X_k$ = the independent variables, which primarily are counts and estimates from the state UI system, sample estimates from the CPS, and independent population estimates

$B_0 - B_k$ = the coefficients of the independent variables, where B_0 is the intercept or constant term

---

*The operation of the state UI system results in a wide array of data concerning employment and unemployment. All of these data are available on a statewide basis, while only a few key items are available at the county level. These data generally subscribe to the official concepts and definitions of, and are highly correlated with, total employment and unemployment. Nevertheless, important deviations do exist, particularly in terms of the UI eligibility requirements among the states. In short, these data tend to correspond with the trend, cycle, and seasonal movements of any labor market area, and, hence, are useful variables for predicting total employment and unemployment for that market.

e = the model's random error term; in these models, this term represents the combined effects of all missing independent variables and measurement error in the dependent and independent variables

(Note: In this report, upper case letters will be used to show true parameters, and lower case letters are used for estimated parameters.)

To estimate the coefficients of our model using the traditional process of ordinary least squares (OLS), we require that the variance of the unknown errors be constant across all observations and also be pairwise uncorrelated. In other words, to use OLS, we must assume that the errors have (1) a constant variance, $\sigma^2(e_i) = \sigma^2$; and (2) a zero covariance, $\sigma(e_i, e_j) = 0$, for all i, j; $i \neq j$. Since the dependent variables in our models are socioeconomic, time-ordered sample statistics, we found that assumption (1) was violated in the cross-sectional models and assumption (2) was violated in all models. In theory, we could still use OLS to estimate the coefficient of the models, if we could first transform the observations so that the errors would meet the required assumptions. Such a two-step process is often referred to as weighted least squares (Draper and Smith 1981), where the weights are the estimated autocorrelation coefficients, and the CPS SMSA sample variances. Since the weights themselves were estimated, the estimation process actually used is more appropriately referred to as estimated generalized least squares (EGLS) following Judge et al. (1980).

5. STATE UNEMPLOYMENT MODELS

One of the main objectives of this research was to find a separate model for each state for predicting employment and unemployment that would be based primarily on data for that state. An important advantage of individual models is that each can be tailored to more reflect local labor market conditions than does the Handbook method.* Additionally, we might expect individual models to be more readily understood, and

---

*The Handbook method was introduced in 1960 by the U.S. Department of Labor as a cost-effective substitute for probability sampling for making state and area estimates. It was ingeniously designed to make maximum use of local administrative data from the state UI system and statistics from the national household survey, the only sources of current unemployment data available at that time. The estimating framework is often described as a "building block," because counts from the UI system are inflated using relationships based on national sample data and are then aggregated to derive the total estimate. All states and areas use the same inflation factors; no adjustment is made to account for the wide differences in state UI laws. (See footnote 1.)

hence, accepted by state technicians and the general public than a model based on, say, a pool of cross-sectional time series data for all nondirect-use states. (Models were developed for both employment and unemployment. Space limitations, however, allow for a description of only the unemployment models.)

5.1. Unemployment Rate

Total unemployment is often separated in to "experienced" unemployed, which consists of unemployed job losers and job leavers, and "entrant" unemployed, which consists of unemployed new entrants and reentrants. The former group of people were employed immediately before their current spell of unemployment, while the latter group were not, at least not on a full-time basis. As a consequence, members of the former group might be eligible for unemployment benefits, while members of the latter most likely would not. This is a very important fact, since the primary source of data for our predictor variables is the large body of administrative data resulting from the operation of the state UI systems. Hence, specifying a behavioral model that would reflect the four previous categories would be impractical, because our major source of data would contain sparse information at best on two of the variables. Therefore, as the basis for the unemployment models, we chose the identity

$$CPSU = EXP + ENT \tag{2}$$

where CPSU = total unemployment
       EXP = experienced unemployed
       ENT = entrant unemployed

This dichotomy is a logical economic separation of workers, and, more importantly, focuses attention on the known data limitations problem at the state level. The variables for the model, then, were selected to serve as proxies for these two groups of unemployed workers. (Separate models were not developed for EXP and ENT because the state sample data are too unreliable for that purpose.)

   We chose to specify the dependent variable in the state unemployment models as a rate instead of a level because (1) the rate is the more important economic variable; (2) ratios can be dynamic interaction variables; and (3) pooling of state regression equations, if considered, would be more tractable with rates than levels. Owing to the specification of the dependent variable, the independent variables were also specified as rates. Our first explanatory variable was the state "insured unemployment rate," since that variable corresponds to the variation in the pool of experienced workers who have lost their jobs and are subsequently receiving unemployment compensation. (Of course, some job losers and most job leavers are not eligible for unemployment compensation.) Obviously, we would expect such a variable to be highly correlated with, and hence, explain a large portion of the variation in the total unemployment rate, especially during business cycles. Such was not the case in many states. We were largely unsuccessful in identifying additional UI statistics from the available stock that might help explain more of the expected variation in the total rate caused by EXP, but did find that cyclically sensitive industry employment data, properly specified, served much the same purpose in many states. Hence, an employment ratio variable was added to the model. Together, these variables explained a significant portion of the variation in the total unemployment rate. The next step entailed identifying local statistics that would account for the variation in the rate caused by ENT, and, hence, further reduce prediction error.

Eligibility rules for UI require that a worker's earnings meet some minimum dollar amount, or that a worker's qualifying spell of employment meet a minimum number of weeks. [For a good discussion of the UI system, and the strengths and weaknesses of UI data, see Blaustein (1979).] Entrants, by definition, either have had no full-time work experience or have had no recent labor force attachment, and, hence, are not generally eligible for benefits. As a consequence, we were unable to find local data in the UI system that would adequately account for the variation in the total rate caused by ENT. Use of seasonal dummy variables to account for entrants was largely unsuccessful in the majority of states, probably because of the extent of sampling error in the dependent variable. An examination of national and regional entrant unemployment rates from the CPS showed that the seasonal patterns of entrants were very similar cross-sectionally. Hence, we used the national CPS entrant unemployment rate in all states as a predictor variable to account for (some of) the variation in the total rate that would be caused by the ENT component.

Following our earlier notation, the state unemployment rate regression models for any time period, t, can be generally defined as

$$\text{CPSR}_t = b_0 + b_1 \text{CLMS}_t + b_2 \text{EMP}_t + b_3 T_t + \sum_1^{12} c_i \text{SEAS}_t + e_t \quad (3)$$

where $t = \text{time} = 1, n$; where $n = $ the number of monthly observation in the sample

CPSR = CPS monthly state unemployment rate

CLMS = state insured unemployment rate, defined as the count of claimants certifying to a week of unemployment for the week including the 12th of the month divided by the estimate of the number of persons on payrolls in nonagricultural establishments (hereafter referred to as "CES" employment) for the same state and month

EMP = a ratio or index based on state CES industry employment, for example, construction employment divided by total employment

T = a linear trend variable; $t = 1, n$; where n = number of months in sample period, so that $T_1 = 1$, $T_n = n$

SEAS = a seasonal variable defined as the interaction between a $(0, 1)$ dichotomous variable (D) and the CPS national monthly entrant unemployment rate (USENT), where in January, $D_1 = 1$, and $D_2 = \cdots = D_{12} = 0$; hence, $D_1 \text{USENT}_t = \text{USENT}_t$, the January value for USENT at period t, and $D_2 \text{USENT}_t = \cdots = D_{12} \text{USENT}_t = 0$, etc.

$b_0 - b_3$, c_i = the model coefficients

e = the model stochastic error term

The entrant variable was specified as 12 seasonal variables, instead of a single time series variable, in order to strengthen seasonality in the predicted series. (Additional testing using regional CPS entrant rates is not yet complete.) The independent variable CLMS is in all state models, even though the variable has little explanatory power in a number of the state models. With the exception of the trend variable, all states have the same model variables. The specification of the employment variable, however, varies by state. There are two possible explanations for the linear trend variable: (1) there has been a secular rise in unemployment during the short sample period that has not been captured by the other predictor variables, or (2) the marked, progressive divergence between the insured and CPS unemployment rates which occurred during the sample period (Burtless, 1983) has been captured by the trend variable.

5.2. Measurement Error in Predictor Variables

The use of sample data as independent variables creates the error in variables problem. Judge et al. (1980) and Pindyck and Rubinfeld (1976) discuss the problem and offer solutions. It follows that the coefficients in our models are both biased and inconsistent. Had we omitted such a variable from the model, however, we might have encountered a more serious problem from a grossly misspecified model. We knew that the values of the coefficients had no intrinsic economic meaning for our purposes. More importantly, testing indicated clearly that the inclusion of a variable with measurement error substantially improved the predictive accuracy of the models. Wonnacott and Wonnacott (1970) suggest that errors in variables is most important if we are concerned with estimation of parameters. However, if we are concerned primarily with prediction, then the "OLS estimator . . . would be the optimal predictor." As we stated earlier, the main objective of these models is to predict estimates with minimum annual MSE. Therefore, we did not consider error in variables to be a significant model development problem, and proceeded as though all of the predictor variables and the error term were independent.

6. AREA MODELS

As we mentioned previously, the type of method that can be used to derive labor force estimates for local areas is a function of the availability of reliable data for each area. For programmatic purposes, the United States has been divided into approximately 2500 mutually exclusive and exhaustive labor market areas (LMAs), the majority of which are single counties. Many of the more populous LMAs are SMSAs. To date, testing of area equations has been limited to a subset of SMSAs, in general, the larger 200 SMSAs for which CPS data are available (hereafter referred to as "CPS LMAs"). These are the more populated LMAs, and, hence, are considered major LMAs for this work.

Ideally, we would like to have a separate regression equation for each of these LMAs for estimation purposes, as we have for each nondirect-use state. However, intercensal CPS sample estimates for most of these small areas do not exist, and the CPS data that are available for small areas are of questionable reliability. Therefore, a separate-equation approach is not possible for most of these LMAs. Fortunately, there are a number of alternative statistical methods that could be used in lieu of an individual regression model for deriving reasonable estimates for each substate area. [For a description of several alternative methods, see Purcell and Kish (1979)].

6.1. Unemployment

The equations for developing substate unemployment estimates build upon the research study of Mathematical Policy Research (MPR), which was conducted in 1980 under a contract with BLS. Following the seminal work of Ericksen (1974), MPR developed a model using the "regression-sample" approach that combined monthly time series data on the equation's variables, cross sectionally, to form a pool of observations for estimation purposes. MPR demonstrated that pooling of CPS area data, in conjunction with data for the independent variables, could be used to estimate the coefficients of a single equation that would subsequently be used to derive an unemployment rate for each area in the pool, even though the CPS data for an individual area were not reliable enough to be used for the same purpose.

The BLS area model is a refinement of the MPR model. The dependent variable and most independent variables were specified as rates to mitigate problems that arise when data for areas of substantially different population sizes are pooled. The independent variables selected for the model clearly reflect the paucity of local explanatory variables that can be used for estimation purposes. The model for any area is generally defined as (for simplicity, the subscripts for time are omitted)

$$\text{CPSR}_i = B_{0i} + \sum_1^k B_k X_{ik} + \sum_1^{12} C_j Z_j + e_i \tag{4}$$

where $i = 1, m$, where $m = $ the number of areas in the pool
$\text{CPSR}_i = $ the CPS total unemployment rate for the ith SMSA in the pool;
$B_{0i} = $ intercept term that is specific to the ith area and is defined as a $(0, 1)$ indicator variable with a value of one for ith area and zero for all other areas
$B_k, C = $ the slope parameters
$X_{ik} = $ the "k" (continuous) independent variables, consisting of an insured unemployment rate (CLMS), a ratio of CPS employment to CPS population (EMPOP), and the official state total unemployment rate (STUR)
$Z_j = a(0, 1)$ seasonal dummy variable, where $Z_j = 1$ for $j = 1, 6, 7,$ and 0 otherwise
$e_i = $ the model stochastic error

A number of independent variables, including population age ratios, and local CES employment ratios, were tested. Those in the following model performed best overall. Nevertheless, the intercept term was allowed to shift across all of the 200 SMSAs in the pool in order to capture interarea structural differences not accounted for by the independent variables in the model. As specified, this is a covariance model (Pindyck and Rubin-feld, 1976), or more specifically, a least squares with dummy variables (LSDV) model (Dielman, 1983), with $m - 1$ dummy variables to account for interarea differences. A number of alternative specifications were examined, but all were rejected since none improved predictions. Weighted least squares was used in the coefficient estimation process to account for heteroscedasticity caused by the use of CPS SMSA data. In recognition of the autocorrelation among the residuals, a variation of Goldberger's (1962) best linear unbiased predictor (BLUP) was incorporated in the estimating regression function, which is generally defined for the ith area for time period t as

$$\widehat{\text{CPSR}}_{it} = b_{0i} + b_1\text{CLMS}_{it} + b_2\text{EMPOP}_{it} + b_3\text{STUR}_t + \sum_{1}^{12} C_j Z_{jt} + \hat{p}_i \hat{e}_{it}$$

$$(5)$$

where

$$\hat{e}_t = \sum_{k=0}^{3} \hat{e}_{t-k}/4 \qquad\qquad (6)$$

and

$\hat{p}_i$ = the estimated autocorrelation coefficient for the ith area

The estimated time series component, pe, is based on a four-month average to reduce random error in the predicted estimates, and reduces prediction error when the residuals are autocorrelated. Again, we have an error in variables problem with this specification because we are using a state variable that is measured with error. The trade-off discussed previously with the state models warrants use of this variable in the area model as well.

7. TEST OF MODEL OBJECTIVES

7.1. Annual Test

The primary measure for testing the performance of each model is the magnitude of the estimated errors. If we followed standard econometric procedures for variable selection as described in Draper and Smith (1981), our initial decision rule would be to choose the model that minimized the magnitude of the estimated errors inside the sample period, since that model would produce estimates nearer to the systematic portion of the CPS than the next best model that we could specify. However, as they suggest, such a model may not produce estimates outside the sample period with minimum errors, especially if the correlative structure among the variables has shifted significantly in the subsequent estimation period.

The sample period we used for testing and model development purposes was 1976–1983 for states and 1978–1983 for LMAs. As we know, there were back-to-back recessions during the latter one-half of these short sample periods. In addition, the sample design was changed in a number of states, and large shifts occurred in two key predictor variables. Fitting a model to such a dynamic set of "unplanned" data could easily result in a model that is misspecified, or else in coefficients that could shift significantly in subsequent periods, when new data are added to the sample. We wanted to avoid falling victim to the "optimism principle," which Picard and Cook (1984) explain can occur from assessing the predictive power of a model on the basis of some in-sample test statistic such as MSE. Therefore, we did not consider in-sample testing alone sufficient for model selection purposes, nor the primary level of evaluation testing.

As we stated previously, a major objective of these models was to produce estimates outside of the sample estimation period with minimum error relative to the CPS. Since the CPS annual average estimates are the official estimates for the 39 nondirect-use states and the District of Columbia, then the most crucial test for each model focused on the magnitude of annual estimation errors relative to the CPS, outside of the sample period used to estimate the coefficients of the model. Therefore, the decision rule that we followed was to choose as the better models those that most efficiently used local data to predict estimates with minimum annual estimation errors, or relative mean square (RMS) errors, in the year immediately following the sample period. This is a strong test of the predictive performance of each model.

As we mentioned previously, the national economy suffered two business cycles since 1976, which is the beginning of our sample period. A reasonable argument against selecting the better models on the basis of a single outside-sample test is that such a specification might not outperform alternative specifications in subsequent periods. As a consequence, we examined the performance of each model using four successive test periods covering 1980–1983. Predictions for 1980 were based on the sample period 1976–1979; predictions for 1981 using the sample period

1976–1980, etc. The models discussed in this report are those that generally produced estimates with the lowest average RMS error over the entire 1980–1983 test period.

7.2. Monthly Test

Next, we tested the best model selected from the preceding ordering process to determine if it produced estimates reflecting local seasonal patterns and business cycle movements. This also was an important model objective. Such testing is difficult because no local data exist that can be used to test directly for these effects in the predicted series. State and SMSA CPS monthly data could be used for this purpose, but, because of large sampling error in the monthly estimates, use of such data would constitute a very weak test. Other correlated local data can be used indirectly to detect general correspondence with these patterns, but not to conduct statistical tests. Therefore, we could only conduct tests on the predicted series itself and infer whether it had attributes that were acceptable. For this testing, we used traditional time series analytical techniques.

In theory, a time series can be decomposed into three components: (1) trend-cycle, (2) seasonal, and (3) irregular. Other things being equal, we can assume that a correctly specified model would produce estimates that correspond to local seasonal patterns and business cycle movements, since the dependent and independent variables in that model are based on local data that supposedly are highly correlated with those patterns. However, some of the variables in our models contain measurement error, in particular, sampling error in the dependent variable. As a consequence, the predicted series from this model could contain an abundance of random error that would obscure the systematic patterns which we are attempting to measure. At this point, we need to restate methodological objective number three to say that our statistical model should "produce a time series with a minimum of noneconomic movement, as well as capture the underlying economic mechanism." Of course, we can expect random shocks to the system, but these should occur infrequently. One test, then, would be to measure the extent of "white noise" or random error in the predicted series. The statistics we used for this test included the standard deviation of percentage month-to-month change, the relative proportion of the irregular component of the series from the Census X-11 variant seasonal adjustment program (Shiskin et al., 1967), and the autocorrelation function of the series for a one period lag. (We expect to further analyze each series at a later date using spectral analysis.) Obviously, evaluation of model performance using these statistics is quite judgmental, somewhat akin to "the proof of the pudding. . . ."

7.3. Area Models

Testing of the area models was less precise because the reliability of the CPS data for a specific SMSA is generally insufficient to evaluate model performance for that area. Hence, we used the aggregate weighted mean, MSE, and cumulative frequency distribution for all areas in the pool as indicators of model performance.

Testing the "expected" within-state distribution of model estimates using data from the 1980 Census is far from straightforward. Many people are unaware of the sampling and nonsampling differences between the CPS and Census that make interpretation of test results difficult. As of this writing, we are assessing such test results.

7.4. Test for Model Stability

We have suggested that the short, turbulent nature of our sample period could easily result in a misspecified model or unstable model coefficients. Stability testing falls under the umbrella of model validation. Snee (1977) discusses cross-validation or data splitting as a means of assessing the predictive performance of a model as well as coefficient stability. Using this technique, the overall set of data is split using an algorithm into two parts: (1) estimation data and (2) prediction data, with the latter set of data used to test a model built on the former. The out-sample testing we described in Section 7.1 is an abridged version of data splitting suggested by Geisser (1975), in which the sample data set of N observations is partitioned into a number of subsets, using $N - n$ observations for estimation, where $N > n$, and n observations for testing. For our purposes, an appropriate means of splitting the data is by time, as is discussed in Draper and Smith (1981) for testing longitudinal data, since our models are clearly temporal in application.

As we indicated previously, we defined the test period as $n = 12$ consecutive observations corresponding to a calendar year. To date, testing has been limited to only a subset of the possible time-related partitions. Eventually, we will estimate the parameters of our "best" model(s) using all $k - 1$ subsets of N monthly data points, where $k = N/12$, and then will predict the monthly estimates for the kth year. Performance will be judged on the basis of overall RMS error and coefficient stability.

Based on our limited testing, we have observed that the model that performed best over the four-year test period did not always perform best in each test year. This is indicative of a model that has not captured all of the variation or information in the dependent variable. We expect to continue model assessment whenever a full year of new data are available, and will change a model specification whenever warranted by test results.

8. ADJUSTMENT TO REDUCE RANDOM ERROR IN PREDICTED SERIES

We hypothesized at the outset of this study that the extent of sampling error in the CPS dependent variable in many states, other things being equal, could make it very difficult to find a model that would produce a series suitable for economic analysis. In addition, the initial time series tests showed that our use of the BLUP predictive model to reduce annual estimation error also had increased random error in the monthly estimates, since the autoregressive errors reflected sampling error in the CPS. As a consequence, we considered various options to reduce the magnitude of random error in the predicted estimates in order to meet our secondary model objective. One option was to smooth the time series component in the BLUP predictive model using a centered or noncentered moving average. An alternative option was to adjust the monthly predicted estimates, derived using only the structural component of the model, using a noncentered moving average ratio correction (MARC) procedure identical in form to the procedure that is currently being used to correct for systematic error in the Handbook estimates. This latter option had the effect of reducing the annual outside-sample prediction error. Based on testing to date, a six-term MARC for unemployment and a four-term MARC for employment appear to perform the best, based on both the annual and monthly tests. In all cases, the predictive models that where tested in conjunction with the MARC procedure did not include the autoregressive term.

REFERENCES

Blaustein, S. (1979), "Insured Unemployment Data," Background Paper No. 24, National Commission on Employment and Unemployment Statistics, Washington, DC.

Burtless, G. (1983), *Why is Insured Unemployment So Low?* Brookings Papers on Economic Activity, Brookings Institute, pp. 225–283.

Czajka, J., and Carr, T. (1981), *The Application of Econometric and Statistical Models to the Estimation of State and Local Area Unemployment*, MPR, Washington, DC.

Dielman, T. E. (1983), "Pooled Cross-Sectional and Time Series Data: A Survey of Current Statistical Methodology," *The American Statistician*, **37**, 111–122.

Draper, N. R., and Smith, H. (1981), *Applied Regression Analysis*, Wiley, New York.

Ericksen, E. P. (1973), "A Method of Combining Sample Survey Data and Symptomatic Indicators to Obtain Population Estimates for Local Areas," *Demography*, **10**, 137–160.

Ericksen, E. P. (1974), "A Regression Method for Estimating Population Change of Local Areas," *Journal of the American Statistical Association*, **69**, 867–875.

Geisser, S. (1975), "The Predictive Sample Reuse Method with Applications," *Journal of the American Statistical Association*, **70**, 320–328.

Goldberger, A. S. (1962), "Best Linear Unbiased Prediction in the Generalized Linear Regression," *Journal of the American Statistical Association*, **57**, 369–375.

Judge, G. G., Griffiths, W. E., Hill, R. C., and Lee, T. (1980), *The Theory and Practice of Econometrics*, Wiley, New York.

Picard, R. R., and Cook, D. R. (1984), "Cross-Validation of Regression Models," *Journal of the American Statistical Association*, **79**, 575–583.

Pindyck, R. S., and Rubinfeld, D. L. (1976), *Econometric Models and Economic Forecasts*, McGraw-Hill, New York.

Purcell, N. J., and Kish, L. (1979), "Estimation for Small Domains," *Biometrics*, **35**, 365–384.

Shiskin, J., Young, A., and Musgrave, J., (1967), "The X-11 Variant of the Census Method II Seasonal Adjustment Program," Technical Paper No. 15, Bureau of the Census, U.S. Department of Commerce, Washington, DC.

Snee, R. D. (1977), "Validation of Regression Models: Methods and Examples," *Technometrics*, **19**, 415–428.

Wonnacott, R. J., and Wonnacott, T. H. (1970), *Econometrics*, Wiley, New York.

Sensitivity of Small Area Estimators to Misclassification and Conceptual Differences of Variables

E. B. Dagum, M. A. Hidiroglou, and M. Morry
Statistics Canada

ABSTRACT

Wages and salaries in small areas are estimated by applying relationships obtained from a sample file to an auxiliary variable available on a universe file. This study investigates the impact of classification and conceptual differences between the two files on the performance of the following small area estimators: count synthetic, ratio synthetic, regression count, and regression ratio. A Monte Carlo simulation is carried out on the data from Ontario to assess the sensitivity of the estimators to these discrepancies.

1. INTRODUCTION

In recent years there has been an increasing demand for statistical information at subprovincial levels. Most of the surveys at Statistics Canada were designed to produce reliable estimates at the province level but not at a lower level of disaggregation. To answer the need for small area statistics without increasing the cost and response burden associated with the larger sample size necessary for producing such statistics through a survey, attention was directed toward the possible use of administrative records. This paper discusses the problems encountered with the reconciliation of data from two administrative files at Statistics Canada and Revenue Canada considered as possible sources for producing small area statistics, related to business income and wages and salaries. It also investigates the impact of conceptual differences and discrepancies in industrial classification between the two files on a set of small area estimators. This latter analysis is complemented by a simulation study carried out on unincorporated business data in the accommodation industry from Nova Scotia and Ontario to produce wages and salaries estimates at the Census Division level.

In 1971, Statistics Canada (STC) was given access to income tax data for the purpose of statistical analysis through the Statistics Act. Tax data for unincorporated businesses (T1) (the scope of this present study) have been transcribed by Statistics Canada since 1973 on a sample basis to produce a file known as the COMBINED-MASTER. It contains a 25% sample of the unincorporated universe with gross business income (GBI) between $25,000 and $500,000 and a 100% sample for tax filers with gross business income over $500,000.

The second administrative file considered in this study is created at Revenue Canada (RC) by transcribing certain tax items for all unincorporated businesses with income over $25,000. The original purpose of this file called COMSCREEN was to serve as a tool in Revenue Canada's auditing procedure. The two files contain a number of economic variables that are comparable in concept. These are sales (known as GBI at STC), capital cost allowance (depreciation at STC), net profit, filer's share of net profit for filers that are involved in partnership. The COMBINED-MASTER has two additional economic variables that are not transcribed on the COMSCREEN file. These are wages and salaries (W&S) and inventories and assets. Thus one file (COMSCREEN) is more complete in coverage but contains less information, while the second file (COMBINED-MASTER) has more variables of interest but only on a sample basis.

In a previous study (Dagum et al., 1984), the authors investigated the possibility of producing wages and salaries statistics for small business in small areas using the two files. They found that a strong linear relationship existed between W&S (available on the COMBINED-MASTER on a sample basis) and GBI (sales) (which was present on COMSCREEN for all T1 tax filers) at the major division industrial breakdown by province. It was necessary to introduce a square root of GBI transformation to the data to make the residuals of the regression homoscedastic. This transformation led to a ratio-type estimation. Applying the ratio obtained from the sample file to sales on the universe file produced improved estimates of wages and salaries at the small area level when compared to the ones obtained by simply blowing up the sample at the small area level. This estimation technique is otherwise known as ratio-synthetic estimation (SYN/R).

Other small area estimators included in that study were the post-stratified estimator (POST), the count-synthetic estimator (SYN/C) popularized by Gonzalez (1973), and two estimators developed by Särndal (1981) and Särndal and Råbäck (1983) to correct for the bias introduced through synthetic estimation called regression-count (REG/C) and regression-ratio (REG/R).* A simulation study carried out on small business data from Nova Scotia indicated that in terms of mean square error (MSE) and especially in small domains, SYN/R was the most efficient followed by REG/R. In terms of bias of the estimates, the REG/R performed best, that is, it produced estimates with minimum bias.

After the completion of the simulation and the analysis of the results, there were several areas that needed further investigation, such as:

(a) Is it possible to cut down on the bias introduced by the ratio-synthetic estimator without considerably sacrificing efficiency as was the case with the regression-ratio estimator?

(b) Does the ranking of the estimators change when moving from a smaller province (Nova Scotia) to a larger province (Ontario)?

Although the previous study pointed out that there were discrepancies in the industrial classification and in the concepts of variables present on the two files, these discrepancies were ignored in the estimation procedure. The simulation used only STC SIC codes and variables, for example, in building the simulated Revenue Canada file it was assumed that the RC SIC codes were identical to those on the Statistics Canada file and that the sales entry of each record on COMSCREEN was the same as the gross business income entry of the corresponding record on the COMBINED-MASTER, that is, that the concepts of sales and gross business income are equivalent. Since in reality there are discrepancies in SIC codes and variable concepts between the SIC and RC files, the following additional questions arise:

(c) What is the difference in wages and salaries when tabulating by RC SIC code versus STC SIC code?

(d) How much extra error is introduced in the estimation procedure by using RC SIC codes and concepts?

*The formulas for these estimators are given in Appendix A.

(e) Does the ranking of the estimators change due to the added discrepancy in coding and concepts?

The objective of the present study is to answer these five questions.

Section 2 introduces two new estimators and compares their performance to five other estimators based on efficiency, bias, and coefficient of variation measures using a simulation on data from the Nova Scotia accommodation industry. Section 3 evaluates the impact of the size of the province on the ranking of estimators by comparing simulation results from Ontario to those from Nova Scotia.

Wages and salaries are tabulated according to both STC and RC SIC codes in Section 4 to assess the difference resulting from misclassification. Section 5 presents the results from a simulation on Nova Scotia and Ontario data using Revenue Canada classification and concepts. The estimates are compared to those obtained earlier in a similar simulation based on STC SIC codes and concepts. This section also examines the impact of coding discrepancies and conceptual differences on the ranking of the estimators used. The conclusions of the study are given in Section 6.

2. AN EVALUATION OF SMALL AREA ESTIMATORS USED TO ESTIMATE WAGES AND SALARIES FOR UNINCORPORATED BUSINESSES

In order to study the properties of various small area estimators, a simulation was undertaken by the authors in 1984. The simulation mimicked the use of administrative data arising from several sources and their subsequent combination to yield small area estimates. Since the Statistics Canada administrative file had all the required information (if only on a sample basis), it was used as the file for drawing the samples required for the simulation. Five hundred samples of size 429 were selected from the target population of 1678 unincorporated businesses in Nova Scotia. The small areas of interest were major industrial groupings by Census Division.

The estimators used [direct (DIR), POST, SYN/C, SYN/R, REG/C, and REG/R] were evaluated in terms of several criteria, such as (1) relative percentage efficiency, (2) relative percentage bias, and (3) coefficient of root mean square error. The estimator that produced the lowest root mean square error (RMSE) of the estimates was the ratio-synthetic estimator. However, it suffered from the drawback of introducing a relatively large bias. Särndal's REG/R, which came second with respect to mean square error and which was nearly unbiased, on the other hand, generated estimates with relatively high MSE. This present study will consider two new versions of the regression-ratio estimator designed to improve the RMSE of the estimates without significantly deteriorating the bias.

The first of the two modified REG/R [MREG/R(1)] developed by Hidiroglou and Särndal (1984) [see estimator $t_7(ai)$ in Appendix A] gives gradually less weight to the residual correction term since the realized sample take deviates from the expected sample take. This may introduce a small bias in exchange for a reduced variance contribution when the realized sample take is lower than expected.

Another alternative to the REG/R was based on the rationale that given the small domains in question, if the realized sample take is lower than expected, the resulting sample size is so small that the correction term is not reliable at all. Thus, if this is the case, the correction should not be used and the estimator defaults to the SYN/R. If the realized sample take is higher than expected, the correction term should enter with the corresponding inverse weight. For the formula see estimator $t_8(ai)$ in Appendix A.

The estimators MREG/R(1) and MREG/R(2) were used together with the other six estimators from the previous study to produce small area W&S estimates in the Nova Scotia accommodation industry through a 250 sample Monte Carlo simulation exercise. The universe in the simulation consisted of the 86 businesses found in this industry on the COMBINED-MASTER file.

The formulas for the three measures used in the calculation of the estimators, that is, relative efficiency (RE), relative bias (RB), and coefficient of root mean square error (CRMSE), are given in Appendix B. Table 1 summarizes the results of the simulation at the province level for the eight estimators according to the three measures. The numbers in brackets refer to the ranking of each estimator according to the three measures.

The ranking of the original six estimators remained the same, that is, SYN/R is still the best in terms of RE and CRMSE but almost the worst regarding the RB, while the theoretically unbiased DIR REG/C and REG/R lead in terms of bias but not in terms of efficiency or CRMSE. The two new estimators strike a compromise in performance between SYN/R and REG/R. They improve upon REG/R in terms of RE and CRMSE but at the expense of deteriorating the bias, as expected. MREG/R(2) is more efficient than MREG/R(1), it falls short of the efficiency of SYN/R by only 11% as opposed to MREG/R(2) that has an RE measure 20% higher than SYN/R. On the basis of bias, MREG/R(1) outperforms MREG/R(2). It cuts down on the bias introduced by

Table 1. Performance of Estimators in Nova Scotia Accommodation–Using STC Coding and Concepts

Measure		Estimators						
	DIR	SYN/C	SYN/R	REG/C	REG/R	POST/C	MREG/R(1)	MREG/R(2)
RE	1.000	0.369	0.271	0.725	0.425	0.809	0.326	0.302
	(8)	(4)	(1)	(6)	(5)	(7)	(3)	(2)
RB	0.045	0.459	0.361	0.040	0.045	0.336	0.195	0.209
	(2–3)	(8)	(7)	(1)	(2–3)	(6)	(4)	(5)
CRMSE	1.076	0.533	0.403	0.846	0.626	0.747	0.416	0.415
	(8)	(4)	(1)	(7)	(5)	(6)	(2–3)	(2–3)

SYN/R by 46% compared to 42% reduction by MREG/R(2). In terms of CRMSE, the two new estimators rate the same, their measure is only 3% higher than the best performing SYN/R's and they perform much better than REG/R that was 55% worse than SYN/R, based on this measure.

In the final analysis, it can be said that the two new estimators introduced are a definite improvement over REG/R. They are only slightly worse than SYN/R regarding root mean square error and efficiency at the same time they substantially reduce the bias associated with SYN/R. In overall performance MREG/R(2) is preferable to MREG/R(1) to some extent. However, it is also somewhat more biased.

3. THE EFFECT OF THE SIZE OF THE PROVINCE ON THE PERFORMANCE OF THE ESTIMATORS

In order to assess the impact of sample size on the optimality of estimators, a Monte Carlo simulation was carried out on data from Ontario. Because of the costs associated with working on data from such a large province, the simulation exercise was restricted to the accommodation industry only. This yielded an 1874 record universe taken off the COMBINED-MASTER file from which 250 25% samples were drawn for the simulation study.

Table 2 presents the performance measure of the eight estimators at the province level according to the same three criteria that were applied to the Nova Scotia data, with the corresponding rankings in brackets. As expected, the size of the relative bias and root mean square error is smaller in Ontario than in Nova Scotia for each and every estimator.

In terms of efficiency the estimators do not improve on the performance of the direct estimator as much in Ontario as in Nova Scotia, but it has to be borne in mind that the percentage root mean square error of the estimates obtained from direct sample blow-up are much smaller in Ontario than in Nova Scotia (0.571 vs 1.076). These results are not surprising given the difference in sample size between the two provinces. What is of more interest to us is whether the larger sample size affected the ranking of the estimators and in which direction?

Comparing the rankings of the estimators in Nova Scotia and Ontario (the numbers appearing in the brackets of Tables 1 and 2), it is evident that the three top performing estimators, namely, the ratio-synthetic and the two modified regression ratio estimators, continued to stay at first, second, and third place in terms of CRMSE and RE even in Ontario. The increase in the sample size did not affect the low ranking of these

Table 2. Performance of Estimators in Ontario Accommodation–Using STC Coding and Concepts

				Estimators				
Measure	DIR	SYN/C	SYN/R	REG/C	REG/R	POST/C	MREG/R(1)	MREG/R(2)
RE	1.000	0.841	0.352	0.770	0.471	0.810	0.413	0.394
	(8)	(7)	(1)	(5)	(4)	(6)	(3)	(2)
RB	0.031	0.333	0.169	0.022	0.015	0.064	0.051	0.090
	(3)	(8)	(7)	(2)	(1)	(5)	(4)	(6)
CRMSE	0.571	0.340	0.174	0.434	0.265	0.458	0.206	0.202
	(8)	(5)	(1)	(6)	(4)	(7)	(2–3)	(2–3)

estimators according to the relative bias measure either. Similarly, the direct blow-up estimator remained in the last place when moving to a larger province. The most significant change occurred in the performance of SYN/C. While in the smaller province, the domain count combined with the average W&S was relatively adequate in describing the variation in wages and salaries among small areas reflected by a rank 4 in terms of RE assigned to SYN/C; in Ontario this type of estimation procedure dropped to seventh place outperforming only the direct sample blow up in terms of efficiency.

Only minor reshuffling of ranks took place in terms of bias among the first second and third rank estimators and also among the fourth, fifth, and sixth rank ones. Based on RMSE, SYN/C and REG/R switched places in ranking and so did the sixth and seventh ranking REG/C and POST/C estimators.

It can be concluded that although the performance of some of the estimators was affected by the size of the province, SYN/R still proved to be the most successful followed by MREG/R(1) and MREG/R(2).

4. DIFFERENCE IN WAGES AND SALARIES RESULTING FROM DISCREPANCY IN INDUSTRIAL CLASSIFICATION BETWEEN STC AND RC

In the preceding analysis it was assumed that there were no differences in SIC coding and in concepts used by the two agencies. Before introducing this added source of error into the estimation procedure, it is worthwhile to examine the magnitude of discrepancy resulting from tabulating W&S according to RC SIC codes instead of STC SIC codes.

For this purpose the COMBINED-MASTER file was matched against the COMSCREEN file using the tax filer's social insurance number as the matching key to create what will be referred to as the MATCHED file. Only single businesses were included in order to avoid mismatches between the two files. Whenever a match occurred, a new record was created containing all the information from the STC file as well as the industrial classification and the sales entry from the RC file. Information on this MATCHED file was then tabulated according to both STC and RC SIC codes. Table 3 indicates the number of businesses and the total wage bill paid out in the major division industrial grouping as coded by STC and RC.

The ratio of the W&S total according to STC and RC classification is used as the measure of the discrepancy. In 8 out of the 17 industry groupings, total W&S according to RC coding is less than the total

Table 3. Discrepancy in Wages and Salaries due to Coding Differences at the Major Division SIC Level for Canada

Major Division	Number of Units (STC)	Number of Units (RC)	W&S (STC)	W&S (RC)	W&S RC / W&S STC
1. Agriculture	170	237	3,031	3,270	1.079
2. Fishing	8	13	159	177	1.113
3. Logging and forestry	685	630	19,007	17,521	0.922
4. Mining	61	62	1,255	1,139	0.907
5. Manufacturing	2,127	1,294	47,451	29,964	0.631
6. Construction	8,495	8,041	136,607	131,420	0.962
7. Transportation	2,806	3,033	40,317	39,811	0.988
8. Communication	149	213	3,166	4,139	1.307
9. Wholesale	1.023	636	11,382	7,618	0.669
10. Retail	12,603	13,492	158,835	179,881	1.132
11. Finance and insurance	7	14	48	149	3.122
12. Real estate	169	108	2,676	1,666	0.622
13. Business service	337	344	6,900	7,153	1.037
15. Educational service	68	106	1,609	1,913	1.189
16. Health and social	254	329	6,947	8,257	1.189
17. Accommodation	4,039	3,936	84,957	83,105	0.978
18. Other services	3,588	3,822	61,003	63,780	1.046
19. Unknown classification	0	282	0	3,211	–

according to STC coding. The ratio among these eight industries ranges from 0.622 for real estate to 0.988 for transportation, that is, RC figures fall short of STC figures by 37% in the worst case and 1% as the closest agreement. In the remaining industries, RC total wages exceed STC total wages anywhere from 4% (business service) to 212% (finance and insurance).

Table 3 suggests that in certain industries such as manufacturing, communication, wholesale, finance and insurance, real estate, educational service, and health and social service some recoding of SIC on the RC file needs to be carried out before it can be used in any tabulation or estimation procedure. These results basically agree with a previous analysis by the authors (Dagum et al., 1984) that measured the coding differences in terms of tabulated GBI.

5. IMPACT OF MISCLASSIFICATION AND CONCEPTUAL DIFFERENCES ON THE ESTIMATION OF W&S IN SMALL AREAS

With the exception of the direct estimator, all the estimators discussed in this paper make use of some of the administrative information contained on the RC file. SYN/C and REG/C obtain the universe count of businesses per small area from the COMSCREEN, while the SYN/R type estimators use the universe total of sales per small area as it is present on the COMSCREEN file. Those small area estimators that are based on counts will introduce an error into the estimates originating from the miscoding of industrial classification present on the COM-SCREEN file. The estimates from the ratio-type estimators that make use of the relationship between GBI and W&S will not only be subject to misclassification errors, but, in addition, they will be influenced by the replacement of the GBI concept with the RC sales concept in the estimation procedure. The estimator that is expected to be most affected by erroneous information on the COMSCREEN universe file is SYN/R because it relies most extensively on COMSCREEN data. Therefore, the analysis to follow will concentrate on the effect of the discrepancies between the two files on the estimates produced by SYN/R. The results will represent the maximum error introduced into the estimates by any estimator due to the discrepancies between the files.

5.1. Analysis of Errors Introduced to the SYN/R Estimates due to Coding and Conceptual Differences, at the Canada Level

For the purposes of this analysis, it was assumed that the small business records available on the MATCHED file constitute the unincorporated universe. This simplification was not expected to affect the findings at the Canada, province, or industry breakdown level. The ratio of W&S and GBI at the provincial level in each industry was applied to all the Census Division GBI totals within that province and industry to produce W&S estimates at the Census Division industry grouping level. This estimation was carried out twice; first the ratios were applied to GBI using STC SIC codes, then they were applied to sales using RC SIC codes. The resulting two sets of estimates were then compared to the true W&S total at the Census Division per industry grouping level to produce two sets of errors at that level. The errors were calculated according to the following formulas:

$$
\text{error}(ai) = \frac{\displaystyle\sum_{a=1}^{A}\sum_{k=1}^{N_{ai}} \text{W\&S}_{aik}}{\displaystyle\sum_{a=1}^{A}\sum_{k=1}^{N_{ai}} \text{GBI}_{aik}} X_{ai.} - \text{W\&S}_{ai.} \tag{1}
$$

where A = number of Census Divisions
N_{ai} = number of units in Census Division a, industry i

and where in the first estimation (STC), $X_{ai.}$ is total GBI in Census Division a, industry grouping i (STC SIC) and in the second estimation (RC) $X_{ai.}$ is total sales in Census Division a, industry grouping i (RC SIC).

These errors can be summarized at the Canada industry grouping level through several statistics, such as the mean absolute percentage error (MAPE):

$$\text{MAPE}_i = \left(\sum_{a=1}^{A} \frac{|\text{error}(ai)|}{\text{W\&S}_{ai}} \bigg/ A \right) \times 100\% \tag{2}$$

and the total absolute percentage error (TAPE):

$$\text{TAPE}_i = \frac{\sum_{a=1}^{A} |\text{error}(ai)|}{\text{W\&S}_{.i.}} \times 100\% \tag{3}$$

While the MAPE statistic gives equal weight to each Census Division regardless of size, the TAPE statistic takes size into consideration to some extent.

Table 4 compares the errors from the two sets of estimates at the Canada Major Division industry grouping level based on the above two statistics.

Table 4 indicates that based on STC codes and concepts alone, only a few industries show estimates with less than 50% MAPE at the small area level, as attested by the entries in column 1. These industries are fishing, construction, retail, and accommodation.*

Judging by column 2 (MAPE using RC codes and concepts) the errors increase substantially in most industries when the complication of incompatible codes and concepts is added to the estimation procedure. The only industries in which the percentage error resulting from the combined effect of estimation and miscoding is relatively small is again fishing, construction, retail, and accommodation. According to column 3, misclassification and conceptual differences increased the error by a maximum of 37% in these four industries. In a few industries such as mining, wholesale, real estate, and educational services, the estimates actually got

*However, the previous study by the authors (Dagum et al., 1984) indicated that reliable estimates can be produced at Census Division level by the ratio-type estimator by treating industries other than construction, retail, and accommodation as one common industry grouping. These four new industry estimates at the Census Division level can then be added up to yield Census Division data.

Table 4. Comparison of Ratio Estimation Errors using STC and RC Coding—Breakdown by SIC; for Canada

Major Division	MAPE STC (1)	MAPE RC (2)	MAPE RC / MAPE STC (3)	TAPE STC (4)	TAPE RC (5)	(TAPE RC) – (TAPE STC) (6) = (5) – (4)
1. Agriculture	265.8	311.4	1.171	38.8	58.2	19.4
2. Fishing	45.9	0.0	0.000	67.4	0.0	–67.4
3. Logging and forestry	129.8	134.6	1.037	28.2	33.4	5.2
4. Mining	472.4	155.2	0.329	36.8	37.0	0.2
5. Manufacturing	62.5	126.7	2.027	18.9	41.3	22.4
6. Construction	31.5	32.5	1.032	12.3	14.7	2.3
7. Transportation	72.9	96.6	1.325	26.4	32.9	6.5
8. Communication	213.2	362.2	1.698	24.9	53.1	28.2
9. Wholesale	259.5	140.8	0.543	35.9	45.6	9.7
10. Retail	30.7	42.2	1.374	15.5	18.5	3.0
12. real estate	737.1	179.1	0.243	45.9	46.4	0.5
13. Business service	1066.4	1083.4	1.016	34.6	43.6	9.0
15. Educational service	468.4	154.7	0.330	31.6	59.0	27.4
16. Health and social	147.8	174.2	1.179	33.6	42.3	8.7
17. Accommodation	44.9	46.4	1.033	12.1	13.2	1.1
18. Other services	108.8	262.5	2.413	20.2	25.1	4.9

Table 5. Comparison of Ratio Estimation Errors using STC and RC Coding—Breakdown by Province; for Canada

Province	Number of Units	Total W&S (STC)	MAPE (STC)	MAPE (RC)	MAPE RC / MAPE STC	(TAPE RC) − (TAPE STC)
Alberta	2,338	33,673	123.3	134.8	109.4	4.4
British Columbia	3,799	58,399	308.6	119.4	38.8	3.5
Manitoba	1,526	20,902	100.0	125.0	125.0	4.8
New Brunswick	988	13,939	112.8	268.3	255.0	9.1
Newfoundland	560	7,106	123.3	149.3	121.1	8.1
Nova Scotia	1,235	19,451	136.7	212.5	155.0	6.4
Northwest Territories	15	266	42.5	48.2	113.4	5.4
Ontario	14,523	232,912	107.4	123.2	114.7	5.0
Prince Edward Island	164	2,358	101.7	118.5	116.4	2.9
Quebec	10,045	178,298	239.8	263.6	109.9	5.7
Saskatchewan	1,368	17,625	183.1	109.8	54.6	4.4
Yukon	25	348	15.9	25.8	162.2	13.0
Missing province code	3	76	—	—	—	—

closer to the true value when RC codes were used, but the error was still around 150% (MAPE RC), indicating that the estimates are still unsuitable for publication.

According to the TAPE statistic, the use of RC SIC codes and concepts introduced some extra error in the estimates for practically all industries. These errors range from an extra 0.5% of the total wage bill in the real estate industry to an extra 28.2% of the total W&S in the communication industry.

Because of the presence of industries in which the estimation procedure produces extremely large errors, the MAPE statistics indicate unacceptable quality estimates when summing up Census Divisions by provinces (Table 5) with the exception of estimates in Yukon and the Northwest Territories.* The deterioration in MAPE due to using RC coding, ranges from 9% in Alberta to 155% in New Brunswick. In British Columbia and Saskatchewan the estimates move closer to the value where estimation is based on RC industrial classification and concepts, but the errors are still too large to allow for publication.

The preceding analysis gave a good insight into the expected size of errors in small areas all across Canada produced by the SYN/R when using two sets of classification codes and concepts. It did not take into consideration, however, the effect of sampling nor did it deal with the performance of the other seven estimators subjected to RC coding. To obtain this type of information, a simulation study was carried out on small business data from the accommodation industry in Nova Scotia and Ontario.

5.2. Analysis of a Simulation Study on Nova Scotia and Ontario Data to Assess the Effect of RC Coding on the Performance of Eight Estimators

The file used as the universe file in this simulation study contained all the records from the MATCHED file from Nova Scotia and Ontario that belonged to the accommodation industry either according to STC or RC SIC coding. This selection procedure yielded 89 records in Nova Scotia and 1902 records in Ontario from which 250 25% samples were drawn. In the estimation phase whenever the information originated from the universe file, RC coding and concepts were applied. On the other hand, the calculations concerning sample data were all based on STC codes and concepts. Thus the only estimator that was not expected to be affected by

---

*In these latter two provinces there are no businesses operating in the industries characterized by high error estimates.

Table 6. Performance of Estimators in Nova Scotia Accommodation–Using RC SIC Coding and Concepts

					Estimators			
Measure	DIR	SYN/C	SYN/R	REG/C	REG/R	POST/C	MREG/R(1)	MREG/R(2)
RE	1.000	0.403	0.340	0.730	0.449	0.769	0.365	0.353
RB	0.045	0.429	0.411	0.103	0.104	0.386	0.252	0.256
CRMSE	1.076	0.504	0.443	0.856	0.643	0.734	0.440	0.440

Table 7. Performance of Estimators in Ontario Accommodation–Using RC SIC Coding and Concepts

					Estimators			
Measure	DIR	SYN/C	SYN/R	REG/C	REG/R	POST/C	MREG/R(1)	MREG/R(2)
RE	1.000	0.898	0.350	0.781	0.512	0.800	0.455	0.433
RB	0.031	0.359	0.221	0.084	0.081	0.099	0.116	0.149
CRMSE	0.571	0.366	0.225	0.450	0.296	0.454	0.246	0.247

Table 8. Performance of Estimators Using RC Coding Relative to the Performance of Estimators Using STC Coding

						Estimators			
Measure	Province[a]	DIR	SYN/C	SYN/R	REG/C	REG/R	POST/C	MREG/R(1)	MREG/R(2)
RE RC	N.S.	1.000	1.092	1.255	1.007	1.056	0.950	1.119	1.169
RE SIC	Ont.	1.000	1.067	0.994	1.014	1.087	0.988	1.102	1.099
RB RC	N.S.	1.000	0.935	1.138	2.575	2.311	1.149	1.292	1.225
RB STC	Ont.	1.000	1.078	1.308	3.818	5.400	1.547	2.274	1.655
CRMSE RC	N.S.	1.000	0.946	1.099	1.012	1.027	0.983	1.058	1.060
CRMSE STC	Ont.	1.000	1.076	1.293	1.037	1.117	0.991	1.194	1.223

[a] N.S. = Nova Scotia; Ont. = Ontario.

RC coding was the direct blow-up estimator that relied strictly on the STC sample data. The rest of the estimators all used universe file information either in form of business counts or auxiliary variable (sales) totals. Tables 6 and 7 summarize the performance of the estimators in Nova Scotia and Ontario according to the same three measures as before.

To put the information from Tables 6 and 7 into perspective, it is necessary to compare them to the corresponding entries in Tables 1 and 2. The impact of RC coding on the errors of estimation is measured by the ratio of the RC table entries and STC table entries as shown in Table 8. As expected, the direct estimator is the only one that was not affected by RC classification. In terms of RMSE, the SYN/R changed most in both provinces followed by MREG/R(2) and MREG/R(1). The least affected estimates are the ones produced by the REG/C. Errors actually become smaller when universe counts in POST were obtained using RC classification. Concerning bias, the originally nearly unbiased REG/C and REG/R deteriorated the most. The bias associated with the SYN/R increased only slightly.

In terms of RMSE, the impact of RC coding was significantly larger in Ontario than in Nova Scotia. For example, the estimates produced by SYN/R showed 29% larger errors in Ontario than previously as opposed to only 10% larger errors in Nova Scotia. (The ratio of RC and STC CRMSEs is 1.293 and 1.099 in Ontario and Nova Scotia, respectively.) These figures agree remarkably well with the ratio of MAPE measures in the accommodation industry calculated in Section 5.1. According to those calculations, the MAPE statistic in Ontario accommodation when using RC codes and concepts was 0.229 yielding a MAPE ratio of 1.283. In Nova Scotia the corresponding statistic was 0.402 yielding a ratio of 1.084.

This type of agreement is relatively close even at the Census Division level. To illustrate this point, Table 9 lists the SYN/R CRMSE measures from the simulations together with the absolute percentage error as obtained from the MATCHED file using the two sets of codes and concepts in the accommodation industry for 16 Census Divisions in Nova Scotia.

There is a strong positive correlation between the absolute error and CRMSE statistics for both the STC and RC estimates. Similarly, the ratios (columns 5 and 6) show a relatively strong relationship. In general, the CRMSE measure slightly overestimates the absolute percentage error statistic.

The entries in Table 9 are sorted according to the size of the Census Divisions. It can be observed that, in general, the absolute percentage error (and correspondingly CRMSE) decreases with increasing sample

Table 9. Absolute Percentage Error per Census Division Compared to Coefficient of RMSE for SYN/R According to STC and RC Coding, for Nova Scotia

Census Division	Number of Units	$\left\|\dfrac{\text{error}}{\text{W\&S}}\right\|$ STC	SYN/R CRMSE STC	$\left\|\dfrac{\text{error}}{\text{W\&S}}\right\|$ RC	SYN/R CRMSE RC	$\dfrac{\|\text{error RC}\|}{\|\text{error STC}\|}$	$\dfrac{\text{CRMSE RC}}{\text{CRMSE STC}}$
01	1	0.289	0.310	0.237	0.313	0.817	1.009
04	1	0.120	0.164	0.074	0.171	0.618	1.042
18	2	2.796	2.789	2.640	2.789	0.944	1.000
03	3	0.065	0.129	0.231	0.225	3.539	1.744
07	3	0.174	0.203	0.208	0.204	1.195	1.004
14	3	0.327	0.342	0.355	0.341	1.084	0.996
06	4	0.048	0.119	0.087	0.119	1.821	1.000
10	4	0.413	0.427	0.355	0.428	0.860	1.002
15	4	0.599	0.608	0.533	0.608	0.890	1.000
02	5	0.227	0.253	0.168	0.243	0.741	0.960
05	5	0.103	0.151	0.058	0.151	0.562	1.000
08	5	0.259	0.229	0.479	0.465	1.852	2.031
12	5	0.274	0.293	0.379	0.365	1.383	1.245
11	8	0.027	0.113	0.246	0.238	9.153	2.106
17	12	0.092	0.142	0.324	0.310	3.505	2.183
09	21	0.067	0.129	0.014	0.113	0.212	0.875
Mean		0.368	0.403	0.399	0.443		
Ratio of means						1.084	1.099

Table 10. Ranking of Estimators According to Three Measures—Nova Scotia

						Estimators			
Measure	Origin of SIC Code	DIR	SYN/C	SYN/R	REG/C	REG/R	POST/C	MREG/R(1)	MREG/R(2)
RE	SIC	8	4	1	6	5	7	3	2
	RC	8	4	1	6	5	7	3	2
RB	STC	2-3	8	7	1	2-3	6	4	5
	RC	1	8	7	2-3	2-3	6	4	5
CRMSE	STC	8	4	1	7	5	6	2-3	2-3
	RC	8	4	1-2-3	7	5	6	1-2-3	1-2-3

Table 11. Ranking of Estimators According to Three Measures—Ontario

						Estimators			
Measure	Origin of SIC Code	DIR	SYN/C	SYN/R	REG/C	REG/R	POST/C	MREG/R(1)	MREG/R(2)
RE	SIC	8	7	1	5	4	6	3	2
	RC	8	7	1	5	4	6	3	2
RB	STC	3	8	7	2	1	5	4	6
	RC	1	8	7	3	2	4	5	6
CRMSE	STC	8	5	1	6	4	7	2-3	2-3
	RC	8	5	1	6	4	7	2-3	2-3

size, as expected. What is not intuitively obvious, however, is that the extra error caused by RC coding, as measured by the ratio of error statistics, increases as the small areas get larger. This phenomenon could originate from the fact that misclassification by RC is not uniformly distributed over all Census Divisions, and that misclassification is more likely to occur, the bigger the small area. Once the small area reaches a certain size, though, the proportion of misclassified business units per area will become constant and the observed effect of size on coding discrepancy will disappear. This was the case in Ontario where a tabulation corresponding to Table 9 failed to reveal a correlation between size and the CRMSE ratio for Census Divisions with more than 20 businesses in them.

5.3. Effect of RC Coding on the Ranking of Estimators

Having examined what happened to estimates produced by SYN/R, after introducing RC coding and concepts, it is time to assess whether the resulting increase in error had an impact on the ranking of this estimator relative to the others. Tables 10 and 11 summarize the results in Nova Scotia and Ontario.

In terms of efficiency and RMSE, SYN/R is still number one or tied for first place with the two modified REG/R. In effect, the relative ranking of all the other estimators remained exactly the same even after taking into account the misclassification. A very slight shift occurred in terms of the bias among the first five ranking estimators in Ontario and among the first three ranking ones in Nova Scotia.

In conclusion it can be stated that the estimators considered are not especially sensitive to misclassification of industry grouping or conceptual differences of the variables used. The introduced error due to these factors is relatively small, and although they effect different estimators to different degree, they do not influence the overall ranking of the estimators.

6. CONCLUSIONS

1. The two new modified regression-ratio estimators developed are successful in cutting down the RMSE of the estimates of REG/R without increasing the bias by much. These two new estimators can be considered as a good alternative to the ratio-synthetic one.

2. Moving from a small province like Nova Scotia to a large one like Ontario reduces the error associated with the estimates, but does not significantly effect the ranking of the estimators.

3. Using RC coding and concepts introduces added error in the estimates, for most industries. However, in industries where wages and salaries are estimated with relatively low error when using STC codes and concepts (i.e., fishing, construction, retail, accommodation), the added error due to the use of RC coding and concepts is minimal.

4. The simulation study conducted in the accommodation industry where the correlation between W&S and GBI is strong, indicates that it is the SYN/R that is most sensitive to the misclassification and concepts introduced by using the RC file, with the rest of the estimators following closely. Deterioration in quality is more evident in a larger province than in a smaller province. However, the overall ranking of the estimators does not change due to the added error. SYN/R still outperforms the others with the two modified regression estimators ranking as close second–suggesting that these latter two estimators could be used to advantage in producing reliable W&S estimates that are not prone to excessive levels of bias.

APPENDIX A: ESTIMATORS USED

1. Direct Estimator (DIR)

$$t_1(ai) = \frac{N}{n} \sum_{k=1}^{n_{ai}} y_{aik} = \frac{N}{n} y_{ai.}$$

where y_{aik} = W&S value of kth sampled unit in the ith industry in the ath Census Division

2. Post Stratified Estimator (POST/C)

$$t_2(ai) = (N_{ai}/n_{ai}) y_{ai.}$$

where N_{ai} = population domain size
n_{ai} = sample domain size

3. Count-Synthetic Estimator (SYN/C)

$$t_3(ai) = N_{ai} \bar{y}_{.i.}$$

where $\bar{y}_{.i.} = \sum_{a=1}^{A} \sum_{k=1}^{n_{ai}} y_{aik} / \sum_{a=1}^{A} n_{ai}$
A = number of Census Divisions

4. Ratio-Synthetic Estimator (SYN/R)

$$t_4(ai) = (\bar{y}_{.i.}/\bar{x}_{.i.})X_{ai.}$$

where $\bar{x}_{.i.}$ = overall sample mean of GBI in ith industry
$X_{ai.}$ = population total of GBI in the ai domain

5. Regression-Count Estimator (REG/C)

$$t_5(ai) = t_3(ai) + (N/n)n_{ai}(\bar{y}_{ai.} - \bar{y}_{.i.})$$

where $\bar{y}_{ai.} = \Sigma_{k=1}^{n_{ai}} y_{aik}/n_{ai}$

6. Regression-Ratio Estimator (REG/R)

$$t_6(ai) = t_4(ai) + (N/n)n_{ai}[\bar{y}_{ai.} - (\bar{y}_{.i.}/\bar{x}_{.i.})\bar{x}_{ai.}]$$

where

$$\bar{x}_{ai.} = \sum_{k=1}^{n_{ai}} x_{aik}/n_{ai}$$

7. Modified Regression-Ratio Estimator (1) [MREG/R(1)]

$$t_7(ai) = (\bar{y}_{.i.}/\bar{x}_{.i.})X_{ai} + D_{ai}\sum_{\kappa=1}^{n_{ai}} [y_{aik} - (\bar{y}_{.i.}/\bar{x}_{.i.})x_{aik}]$$

where

$$D_{ai} = \begin{cases} N_{ai}/n_{ai} & \text{if } n_{ai}/N_{ai} \geq n/N \\ (N/n)^2 n_{ai}/N_{ai} & \text{if } n_{ai}/N_{ai} < n/N \end{cases}$$

8. Modified Regression-Ratio Estimator (2) [MREG/R(2)]

$$t_8(ai) \text{ same as } t_7(ai)$$

except

$$D_{ai} = \begin{cases} N_{ai}/n_{ai} & \text{if } n_{ai}/N_{ai} \geq n/N \\ 0 & \text{otherwise} \end{cases}$$

APPENDIX B: MEASURES USED FOR ASSESSING PERFORMANCE

1. Relative Bias of the mth Estimator

$$\text{RB}[t_m(ai)] = \frac{1}{A} \sum_{a=1}^{A} \left| \frac{\bar{t}_m(ai) - Y_{ai.}}{Y_{ai.}} \right|$$

$$= \frac{1}{A} \sum_{a=1}^{A} \left| \frac{\bar{B}[t_m(ai)]}{Y_{ai}} \right|$$

$$\bar{B}[t_m(ai)] = \sum_{r=1}^{250} (t^r_m(ai) - Y_{ai.})/250 , \quad m = 1, 2, \ldots, 8$$

2. Relative Efficiency of the mth Estimator

$$\text{RE}[t_m(ai)] = \frac{1}{A} \sum_{a=1}^{A} \left\{ \frac{\overline{\text{MSE}}[t_m(ai)]}{\overline{\text{MSE}}[t_1(ai)]} \right\}^{1/2}$$

$$\overline{\text{MSE}}[t_m(ai)] = \sum_{r=1}^{250} [t^r_m(ai) - Y_{ai.}]^2/250 , \quad m = 2, 3, \ldots, 8$$

$r = $ replication number in the simulation

3. Coefficient of Root Mean Square Error

$$\text{CRMSE}[t_m(ai)] = \frac{1}{A} \sum_{a=1}^{A} \frac{\{\overline{\text{MSE}}[t_m(ai)]\}^{1/2}}{Y_{ai.}} , \quad m = 1, 2, \ldots, 8$$

REFERENCES

Dagum, E. B., Hidiroglou, M. A., and Morry, M. (1984), "The Use of Administrative Records to Estimate Wages and Salaries for Small Businesses in Small Areas," *1984 Proceedings of the Economic Statistics Section*, American Statistical Association, pp. 472–477.

Gonzalez, M. E. (1973), "Use and Evaluation of Synthetic Estimates," *1973 Proceedings of the Social Statistics Section*, American Statistical Association, pp. 33–36.

Hidiroglou, M. A., Morry, M., Dagum, E. B., Rao, J. N. K., and Särndal, C. E. (1984), "Evaluation of alternative Small Area Estimators using Administrative Records," *1984 Proceedings of the Survey Methodology Section*, American Statistical Association, pp. 307–313.

Hidiroglou, M. A., and Särndal, C. E. (1984), *Experiments with Modified Regression Estimators for Small Domains*, Technical paper, Statistics Canada.

Särndal, C. E. (1981), "Frameworks for Inference in Survey Sampling with Applications to Small Area Estimation and Adjustment for Non-response," *Bulletin International Statistical Institute* **49**(1), 292–513 (Proceedings, 43rd Session, Buenos Aires).

Särndal, C. E. and Räbäck, G. (1983), "Variance Reduction and Unbiasedness for Small Domains Estimators," *Statistical Review*, **5**, 33–40.

The Estimation of the Number of Unemployed at the Small Area Level

G. A. Feeney
Australian Bureau of Statistics

ABSTRACT

Unemployment statistics are key social and economic measures needed to develop and assess policy. The primary source of this information has been the Monthly Labour Force (MLFS). Because the MLFS is based on a sample, enormous cost in terms of dollars and respondent burden would be required in order to provide detailed information for specific small areas. Since many labor market changes, as well as many of the government programs that are affected by labor market conditions are regionally based, we decided to examine the possibility of using data collected in the course of program administration as a source of small area information. For unemployment statistics we obtained data collected by the Department of Social Security (DSS) in the dispatch of unemployment benefits. The technique we have used to marry the two sources of unemployment data—DSS unemployment benefit recipients and ABS Labour Force Survey—involves the use of the structure-preserving estimation (SPREE) method of Purcell (1979). The method will be described and evaluated using Population Census data. It will be shown that for this application of interest the SPREE estimates perform very well.

The opinions expressed in this paper are those of the author and do not necessarily reflect the policies of the Australian Bureau of Statistics.

1. INTRODUCTION

Unemployment statistics are key social and economic measures needed to develop and assess government policy. The primary source of this information has been the Monthly Labour Force Survey (MLFS), which provides tabulations of labor force status by age, sex, marital status, broad occupational categories, and other characteristics of the labor force. Because the MLFS is based on a sample, enormous cost in terms of dollars and respondent burden would be required in order to provide detailed information for specific small areas. Since many labor market changes, as well as many of the government programs that are affected by labor market conditions, are regionally based, the Statistical Services Branch of the Australian Bureau of Statistics decided to examine the possibility of using data collected in the course of program administration as a source of small area information. For unemployment statistics we obtained access to the data collected by the Department of Social Security (DSS) in the dispatch of unemployment benefits.

The technique we have used to marry the two sources of unemployment data—DSS unemployment benefit recipients and the ABS Monthly Labour Force Survey—involves the use of an iterative proportional fitting (IPF) algorithm and is comprehensively described in Purcell (1979). The technique will be referred to as the structure-preserving estimation (SPREE) method.

2. PRELIMINARY ANALYSIS

Because the DSS statistics differ somewhat in concept from those of the commonly used MLFS, some attention should first be paid to the comparison of the two data sources. Indeed, to be successful in the melding of these two sources of unemployment data, some linking variables need to be established.

The MLFS definition of the unemployed covers those who were not employed during the survey week and who have actively looked for full-time or part-time work at any time in the four weeks up to the end of survey week. This definition includes, for instance, individuals who choose not to apply for unemployment benefits. It excludes those with part-time jobs. In contrast, the DSS definition includes only those who are registered and eligible for unemployment benefits.

A large proportion of the unemployed population are covered equally by the MLFS and DSS concepts, but there is a subset of persons who are either not unemployed according to the MLFS definition, but are unemployment benefit recipients, or alternatively, are unemployed according the MLFS definition but are not unemployment benefit recipients.

In September to December 1982 the ABS conducted the Special Supplementary Survey No. 5 (SSS5) using the population survey framework. The sample size was approximately one-half that of the usual MLFS (i.e., sample size of 35,000 persons).

The survey collected, among other variables,

(i) Demographic characteristics, for example, age, sex, marital status, country of birth.

(ii) Labor force status (a derived variable), that is, employed, unemployed, not in labor force.

(iii) Unemployed benefit status, that is, whether or not you are currently receiving unemployment benefits.

These variables were collected for every person aged 15 and over who were residing in the selected dwellings.

When dealing with categorical variables, such as the preceding, the most appropriate statistical techniques for analysis is the log-linear model.

In the present context we firstly derived a 0–1 response variable according to whether the two definitions differed or agreed for each individual and considered the following combinations of demographic variables as effects or factors:

(a) Age group (10 levels) and combined sex – marital status.

(b) Combined age group – sex and marital status.

(c) Combined age group – marital status and sex.

(d) Age group, sex, and marital status.

The models we fitted were of the form

$$\log(P_{ij}/(1 - P_{ij})) = \mu + \alpha_i + \beta_j + \epsilon_{ij}$$

where P_{ij} = the proportion of individuals in the (i, j)th cell where the definitions disagreed

μ = an intercept parameter

α_i = the effect for the ith category of the first variable

β_j = the effect for the jth category of the second variable

ϵ_{ij} = the residual or error term

(The extension to a third variable is obvious.) Hence if the α's or β's are positive/negative, then there is a tendency for disagreement/agreement.

The performance of any particular model is measured in two ways. These are:

1. The statistical significance of the parameter estimates (i.e., μ, α, β) is determined by the associated P-value. This determines whether the effect is significantly different from zero. The P-value refers to the probability that, even though the effect was zero, pure sampling variation could have resulted in a test statistic equal to or higher than that observed.

2. The statistical significance of the residual (i.e., ϵ) is also determined by the associated P-value. This determines whether the data fit the model well or not. The P-value has the same interpretation as in 1, and a nonsignificant residual indicates that the model adequately fits the data.

The results are given in Table 1.

Table 1. Test for Appropriate SPREE Marginal Variables

	DF	χ^2	Probability
Model a			
Intercept	1	25	.0001
Age group	9	19	.02
Sex–marital status	3	111	.0001
Residual	25	15	.9
Model b			
Intercept	1	0.01	.9
Age group–marital status	19	38	.006
Sex	1	39	.0001
Residual	17	80	.0001
Model c			
Intercept	1	0.2	.7
Age group–sex	18	76	.0001
Martial status	1	6	.02
Residual	18	64	.0001
Model d			
Intercept	1	0.2	.7
Age group	9	16	.08
Sex	1	45	.0001
Marital status	1	4	.05
Residual	26	91	.0001

All of the variables in these models were significant in explaining the disagreement but only model (a) had a nonsignificant residual, that is, it alone fitted the data well. Indeed it fitted the data extremely well.

The preliminary analysis indicates that the major determinants of the difference between MLFS and DSS definitions of unemployment are age and sex by martial status. The pattern of agreement or disagreement is not smooth over the various age categories. This may be able to be explained by reference to the DSS eligibility criteria.

The linking variables have therefore been established empirically; they are age and sex by marital status.

3. SMALL AREA ESTIMATION

The SPREE technique, as applied in this situation, is briefly described below.

An association structure, obtained from the current DSS survey, is a cross-classification of the number of unemployment benefit recipients by age and by sex—marital status for each small area. This association structure is used to establish the relationship between the variable of interest and the associated variables (age, sex, marital status) at the small area level. We then use the allocation structure, obtained from the number of unemployed in the MLFS cross-classified by age and sex—marital status at the state level, to establish the relationship between the variable of interest and the associated variables at a level accumulated over all the small areas. The time periods relevant to the association structure and the allocation structure are the same and the SPREE technique is applied to adjust for changes or differences in the definition between the two sources. In addition, and this point is extremely important, each subsequent estimation uses a *different* association structure as well as a different allocation structure. Therefore, the assumptions on the interactions change for each time period.

As an aid to understanding the technique we present a graphical description (Figure 1). The association structure, obtained from DSS data, is defined by π_{hig} where h denotes the small area, i denotes the age category, and g denotes the sex—marital status category. The MLFS provides us with the allocation structure; that is, $\zeta_{.i.}$ and $\zeta_{..g}$. We can then invoke the SPREE algorithm, which adjusts the original three-way table to conform to the i margin then adjusts further to conform to the g margin. This procedure is repeated in an iterative fashion until the values in the three-way table converge. Thus, analytical estimates P_{hig} can be obtained which can then be summed over i and g to arrive at the small

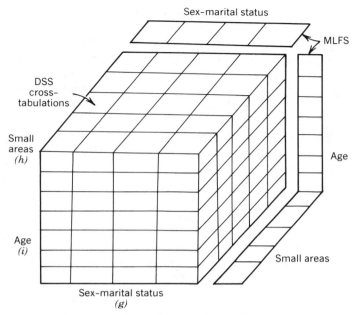

Figure 1.

area estimates of unemployment which, when aggregated, agree with the MLFS estimates.

Of course, estimates at this small area level could have been obtained directly from the MLFS. However, the sample size in each small area would be so small that the sampling error for such an estimate would have been prohibitively high. The estimates from SPREE have a much smaller sampling error, while the modeling introduces a model error or bias that is independent of sample size. It is hoped that the total error is drastically reduced by using the SPREE technique.

4. EVALUATION OF THE SPREE ESTIMATES

The obvious question to ask in relation to this method is, "How good are the estimates?" There is no information at the small area level on the number of unemployed according to the ABS MLFS definition. Without this there is no way of estimating the total error (or its components) for any particular estimate. We can, however, assess the performance if instead of the MLFS definition we use the ABS Population Census definition of unemployed. If the technique performs well for these data, then it is reasonable to assume that it would perform equally well for MLFS data. Therefore, we can use the DSS data for the June 1981 quarter as the association structure and the number of unemployed at the 1981 Population Census (June 1981) by age and by sex–marital status as the allocation structure. The resultant estimates can be compared with the *known* Census data at the small area level.

This evaluation was done for Local Government Areas in four states of Australia (New South Wales, Victoria, Queensland, and Western Australia). These results can be presented in a number of ways; however, we will concentrate on the following summaries.

(i) Scattergrams of SPREE estimate versus Census unemployed.

(ii) Regression equation (correlation coefficient).

(iii) The percentage difference (in graphical form) versus the Census value. This display will detect any outliers.

(iv) The misallocation from using the SPREE estimates rather than the Census values, that is,

$$\text{Percentage misallocation} = \frac{\frac{1}{2} \sum_{\text{all LGAs}} |\text{SPREE–Census}|}{\text{total census unemployment}} \times 100$$

Note that this measure ranges from 0 to 100 and so does not take account of any opportunity costs incurred by a misallocation of funds.

(v) The standard errors implied by the relationship assuming the same form of the model as used for the MLFS standard errors. The theory behind this model is as follows:
Let y'_i be the estimate for the ith LGA and x_i be the true value of the ith LGA. Then

$$y'_i = \beta x_i + \xi_i \tag{1}$$

where

$$E(\xi_i) = 0$$

$$\text{Var}(\xi_i) = \sigma^2 x_i^{2\gamma}$$

Using the data from all LGAs within a state and deleting one observation to ensure independence (since LGA estimates must add to the State total) of the y_i''s, we can estimate β, α and γ. Then

$$\text{Vâr}(y_i') = \hat{\sigma}^2 x_i^{2\hat{\gamma}} \simeq \hat{\sigma}^2(y_i')^{2\hat{\gamma}}(\hat{\beta})^{-2\hat{\gamma}}$$

Thus

$$\frac{\text{MŜE}(y_i')}{y_i'} \times 100 \simeq \frac{100}{y_i'} \{\hat{\sigma}^2(y_i')^{2\hat{\gamma}}(\hat{\beta})^{-2\hat{\gamma}} + \underbrace{(1 - 1/\hat{\beta})^2(y_i')^2}_{\text{Bias}^2}\}^{1/2}$$

So

$$\text{RMŜE}(y_i') \simeq 100[\hat{\sigma}^2(y_i')^{2(\hat{\gamma}-1)}(\hat{\beta})^{-2\hat{\gamma}} + (1 - 1/\hat{\beta})^2]^{1/2}$$

Therefore, using the model we can estimate the RMSE based on the true value and the estimate. In the normal situation of the MLFS we can explicitly calculate the variance of each estimate (using split-halves), without knowing the true value. This relationship is then modeled to smooth out the variance estimates using the following model:

$$\log \% \text{ RSE}(y_i') = a + b \log y_i' + c(\log y_i')^2$$

which implies a model approximately of the form (1). We can therefore compare the RMSE and the RSE obtained from the two methods of estimation—SPREE and the direct MLFS.

In using this comparison we must keep in mind that in practice we would be using an allocation structure obtained from the MLFS which is subject to sampling error. However, since this allocation structure is at the State level, the sampling error is moderate.

The results are summarized in Tables 2 and 3.

Table 2. Measures (ii) and (iv) by State[a]

State	R^2	Correlation Coefficient	Percentage Misallocation
New South Wales	.9602	.9688	8.9
Victoria	.9709	.9778	8.0
Queensland	.9792	.9872	9.3
Western Australia	.9831	.9901	7.4

[a]Allocation structure at state level.

Table 3. Relative Standard Error by Size of Estimate and State[a]

Estimate	New South Wales		Victoria		Queensland		Western Australia	
	SPREE	MLFS	SPREE	MLFS	SPREE	MLFS	SPREE	MLFS
10	86.0	376.1[b]	77.0	339.5[b]	80.3	268.6[b]	77.4	292.4[b]
20	71.7	284.7[b]	63.3	263.9[b]	67.6	210.8[b]	63.7	220.5[b]
50	56.3	193.9[b]	48.8	185.1[b]	52.2	149.4[b]	49.3	148.8[b]
100	46.8	143.2[b]	40.1	139.2[b]	42.9	113.1[b]	40.6	108.8
200	39.1	104.6	33.0	103.1	35.3	84.4	33.4	78.5
500	30.7	67.9	25.6	67.9	27.3	55.9	25.9	50.5
1000	25.6	48.4	21.2	48.7	22.5	40.2	21.3	35.0
2000	21.3	34.1	17.5	34.4	18.6	28.5	17.5	24.1
4000	17.8	23.8	14.5	24.0	15.4	19.9	14.5	16.4
5000	16.8	21.1	13.7	21.3	14.5	17.7	17.7	14.5

[a]Allocation structure at state level.
[b]Estimates of this size are impossible from the MLFS; therefore, the model is extrapolated beyond its natural range.

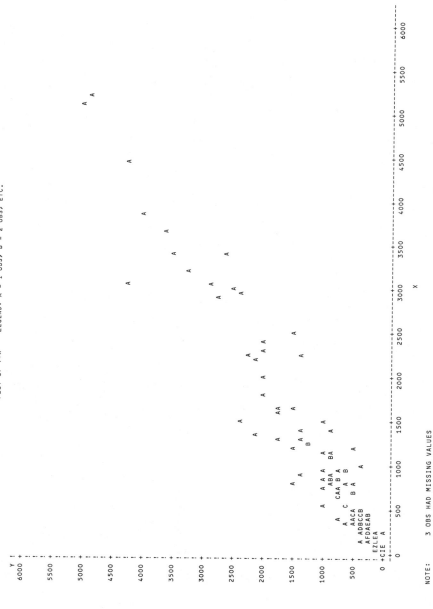

Figure 2. NSW: state level, SPREE vs Census. A = one observation; B = two observations; etc.

207

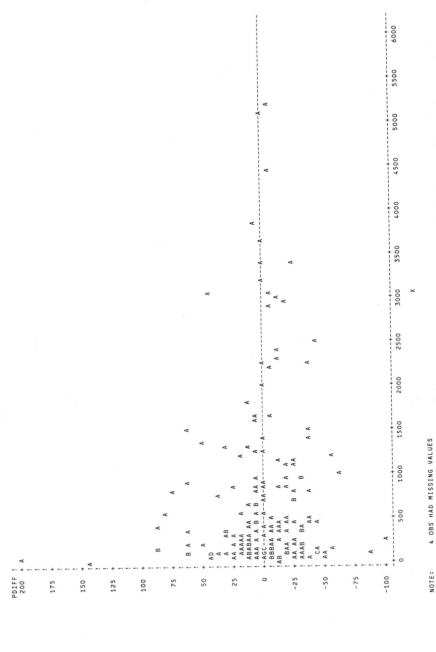

Figure 3. NSW: state level, percentage difference vs Census. A = one observation; B = two observations; etc.

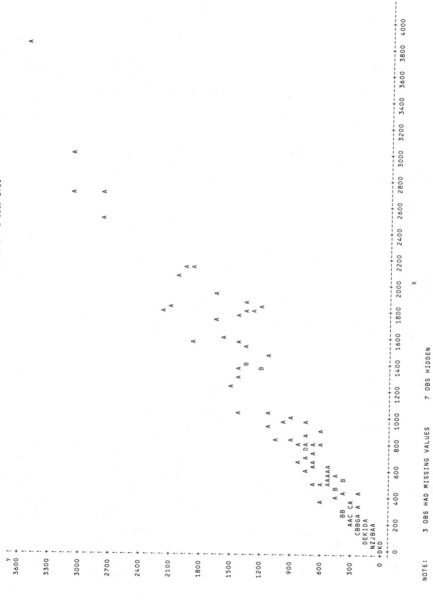

Figure 4. VIC: state level, SPREE vs. Census. A = one observation; B = two observations; etc.

209

------ VIC --------- 20:29 WEDNESDAY, MAY 21, 1986 3

PLOT OF PDIFF*X LEGEND: A = 1 OBS, B = 2 OBS, ETC.

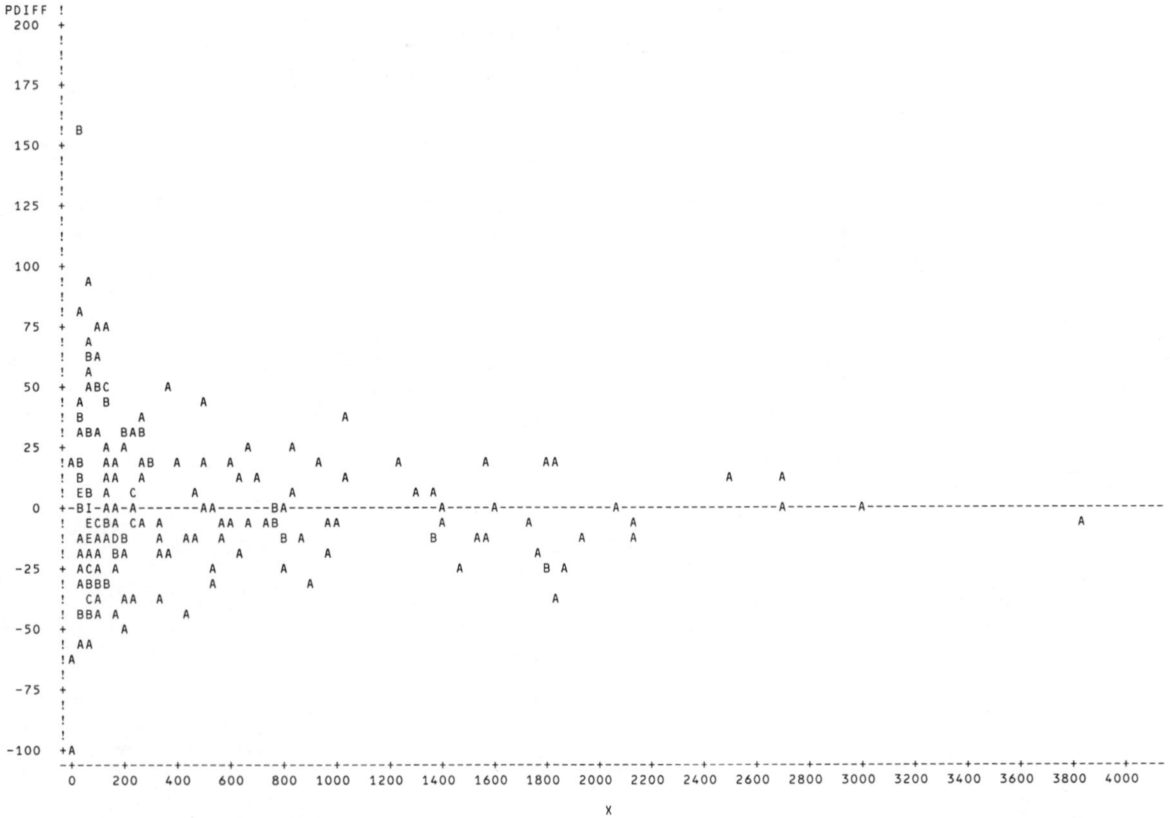

NOTE: 4 OBS HAD MISSING VALUES

Figure 5. VIC: state level, percentage difference vs. Census. A = one observation; B = two observations; etc.

The graphs are quite space consuming so we present only those that show a somewhat unique pattern. We believe that the graphs for states New South Wales and Victoria are quite informative. These are presented as Figures 2–5. They show that there are a number of outliers, especially for New South Wales.

The underestimates are all Sydney (capital of NSW) metropolitan LGAs of a middle class nature. This suggests that for these areas the adjustment for the deficiency of married females on the DSS file (since most married females are ineligible for unemployment benefits) has not been sufficient. Treatment of these outliers will be discussed later. The graphs that are not shown do not exhibit such outlier problems.

5. SOME MODIFICATIONS

We have seen that, overall, the preceding results are encouraging. For NSW, however, there are a few outliers that cause some concern. It would appear that the assumptions implicit in the model used do not hold universally.

Suppose that instead of the allocation structure being at the state level it was constructed at the regional substate level. Are the assumptions more appropriate at this level? We have considered this situation and applied it to the data. The results of this evaluation are presented in the format of the previous section.

The results are summarized in Tables 4 and 5. Again only the graphs for New South Wales and Victoria are presented in Figures 6–9.

We can see that, when the SPREE method is applied at the regional level, the improvement in the results is substantial, especially for NSW.

Table 4. Measures (ii) and (iv) by State[a]

State	R^2	Correlation Coefficient	Percentage Misallocation
New South Wales	.9883	.9909	5.0
Victoria	.9847	.9883	6.2
Queensland	.9860	.9913	8.0
Western Australia	.9946	.9968	5.3

[a]Allocation structure at regional level.

Table 5. Relative Mean Square Error[a]

Estimate	New South Wales		Victoria		Queensland		Western Australia	
	SPREE	MLFS	SPREE	MLFS	SPREE	MLFS	SPREE	MLFS
10	91.1	376.1[b]	68.3	339.5[b]	93.2	268.6[b]	72.6	317.3[b]
20	68.4	284.7[b]	55.9	263.9[b]	71.9	210.8[b]	56.7	235.7[b]
50	46.8	193.9[b]	42.9	185.1[b]	51.1	149.4[b]	40.8	156.4[b]
100	35.1	143.2[b]	35.1	139.2[b]	39.5	113.1[b]	31.9	113.2
200	26.4	104.6	28.8	103.1	30.5	84.4	24.9	81.0
500	18.2	67.9	22.1	67.9	21.7	55.9	18.0	51.2
1000	13.8	48.4	18.1	48.7	16.8	40.2	14.2	35.7
2000	10.5	34.1	14.8	34.4	13.0	28.5	11.2	24.6
4000	8.1	23.8	12.2	24.0	10.1	19.9	8.8	16.8
5000	7.5	21.1	11.4	21.3	9.3	17.7	8.2	14.8

[a]Allocation structure of regional level.
[b]Estimates of this size are impossible from the MLFS; therefore, the model is extrapolated beyond its natural range.

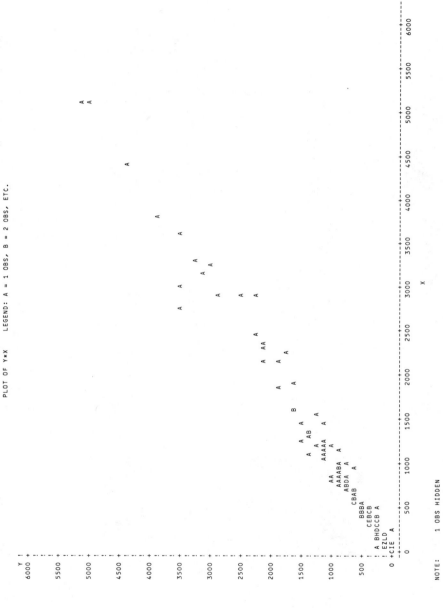

Figure 6. NSW: regional level, SPREE vs Census. A = one observation; B = two observaions; etc.

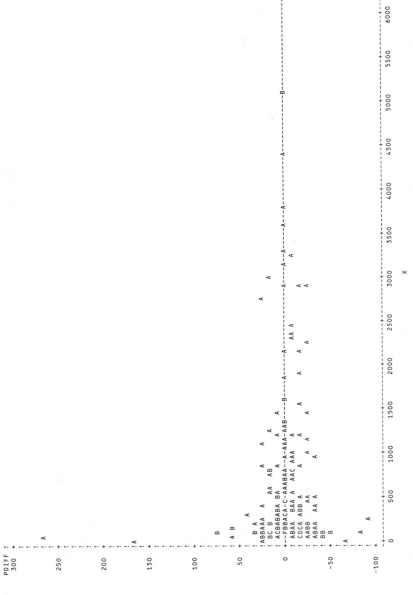

Figure 7. NSW: regional level, percentage difference vs. Census. A = one observation; B = two observations; etc.

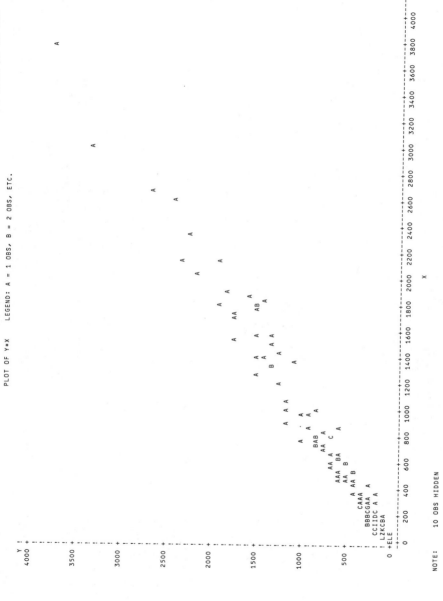

Figure 8. VIC: regional level, SPREE vs. Census. A = one observation; B = two observations; etc.

215

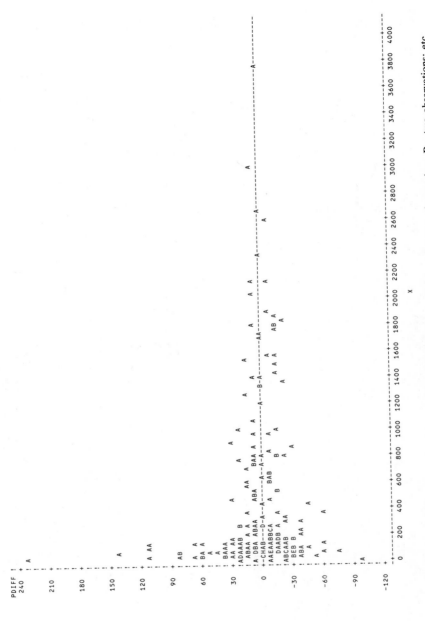

Figure 9. VIC: regional level, percentage difference vs. Census. A = one observation; B = two observations; etc.

216

6. APPLICATION TO THE MLFS

In the preceding two sections we showed that the SPREE method performed well when the allocation structure was Population Census data. If, however, as proposed, the allocation structure was obtained from the MLFS, how well would the SPREE method perform? There are two dimensions to this question.

First, the SPREE method adjusted for the differences between the Population Census and the DSS definition of unemployment. It would not be a too brave assumption that the SPREE method would adjust equally well for the differences between the MLFS and the DSS definition of unemployment. This implies that age, sex, and marital status are also the main determinants of the differences between the DSS and the MLFS data. This is supported by the results given in Section 1. Therefore, the evaluation of the performance of the method is valid.

Secondly, we must examine the measurement of accuracy when the MLFS allocation structure is used. The relative standard errors have been given previously when the Population Census allocation structure was used. This allocation structure is not subject to sampling error, like the MLFS allocation structure. How is this sampling error reflected in the final small area SPREE estimates? This question is difficult to answer theoretically. However, simulations could be conducted to determine the effect of random perturbations in the allocation structure. This may be considered in the future.

7. CONCLUSION

We have shown that it is possible to obtain feasible estimates of the level of unemployment at the LGA level by the use of the SPREE technique to marry together two sources of data.

The preceding results are very encouraging; however, more work now needs to be done to determine the efficacy of this technique in producing a credible time series of unemployment data at the local area level.

ACKNOWLEDGMENTS

I would like to express my thanks to Mr. Brian Harvey and Ms. Megan Werner for their invaluable assistance in the statistical computing undertaken for this paper. I am also grateful to Mrs. Pam Wood for her work in typing this manuscript.

REFERENCES

Purcell, N. J. (1979), "Efficient Estimation for Small Domains: A Categorical Data Analysis Approach," Doctoral Dissertation, University of Michigan. (Copies are available from Dr. G. A. Feeney, Australian Bureau of Statistics, P. O. Box 10, Belconnen, ACT 2615, Australia.)

Purcell, N. J. and Kish, L. (1979). "Estimation for Small Domains," *Biometrics*, **35**, 365–384.

Small Area Estimation Research for Census Undercount—Progress Report

C. T. Isaki, L. K. Schultz,
P. J. Smith, and G. J. Diffendal
U.S. Bureau of the Census

ABSTRACT

The Bureau of the Census is currently investigating the potential use of several strategies for adjusting the census count for small areas. The strategies investigated consist of combinations of regression and synthetic estimation methods. This chapter summarizes background information on the nature of the undercount and its impact on major uses of census data, and describes the available information pertaining to undercount. Adjustment strategies under study are presented together with results obtained to date and plans for future work.

1. INTRODUCTION

In 1980, the U.S. Bureau of the Census reported a census count of 226,549,448 persons on census day, April 1st. No one knows the true number of persons living in the United States. Beginning in 1950, demographic analysis methods were used to estimate the net census undercount. The most recent estimates of net undercount were 1% for 1980 (assuming 2.0 million illegal aliens included in the 1980 census enumeration); 2.8% for 1970; 3.3% for 1960 and 4.4% for 1950.

In 1980, a postenumeration program provided a range of net undercount estimates of roughly a 0.5% overcount to a 2% undercount.

219

Estimates of net undercount for states exhibited a wide range, also. For example, one series of state estimates exhibited a range of a 2% overcount to a 6% undercount. Hence, in addition to variability among net undercount rates for the United States, we also appear to have differential net undercount among states. There is also evidence of differential undercount among race and sex groups. Such differential undercounting has been the basis for a number of court cases in which various jurisdictions, cities as well as states, have sued the Bureau of the Census to adjust the 1980 census counts. To date, the Bureau is not under any court orders to adjust the census counts.

One factor motivating certain jurisdictions to sue for an adjustment of the census is the use of population counts in determining representation in government as well as in determining the amount of revenues received from the federal government. Since congessional representation is determined on a relative population basis among states, if every state experienced the same percentage net undercount, then the allocation of number of representatives to states would remain unchanged whether the census counts or corrected counts were used. Differential net undercount among states would cause a difference in representation. With respect to revenue-sharing allocation to state and substate governments, it has been reported (Robinson and Siegel, 1979) that undercoverage of the income component in the revenue-sharing formula has a greater effect than that due to undercoverage of the population. Nevertheless, undercoverage of the population is still perceived as the cause of incorrect disbursement of revenue-sharing funds and hence the continuation of some jurisdictions to request adjustment of census counts. In addition, other methods of income and population adjustments than that used by Robinson and Siegel may alter their results. Finally, some federal programs base eligibility of jurisdictions on level of total population and allot funds on the basis of other variables. For example, a program for economic development of communities only includes metropolitan cities and urban counties of a sufficient size. The funds alloted to these two types of communities are each based on such variables as population, poverty, and housing overcrowding relative to all such communities. A secondary effect is the use of census population counts in determining ratio adjustment factors in on-going surveys such as the monthly labor force surveys. Estimates of employment status used in disbursing funds would be affected by population undercoverage in this manner.

The research conducted so far has dealt with total population as a characteristic to be adjusted, since it is the first characteristic that is to be produced from the census. The first set of population counts is required by state by the end of the calendar year, while a second set of population counts is required for legislative redistricting purposes a year after census day. Since the basic unit of census tabulation is the census block consisting of an average 100 persons per block, one possibility is to adjust the census block for undercount. Adjustment at this level will then be consistent at higher levels of aggregation and in cross tabulation. We have not concerned ourselves with the problem of adjusting other characteristics, but it is likely that should population be adjusted, housing unit counts, race, age, sex, and other characteristics would require adjustment. Another possibility is to only adjust at higher levels of aggregation. At any rate, the manner in which adjusted counts would be displayed in census publications, if adjustment is implemented, has not been decided upon at this time.

The focus of our small area research so far is in three general directions. The first direction is to look at the results of the 1980 PEP (Post Enumeration Program). The second is to look at demographic analysis results for 1980. The third direction is to use the 1980 census data to simulate and evaluate the performance of potential adjustment methodologies. The remainder of this chapter describes the limitations of the data tools previously mentioned, describes the adjustment methodologies being investigated, and provides the results of our work to date.

2.   DATA USED IN RESEARCH

The data used in our research comes from the 1980 PEP, demographic analysis, and the census. Each data source has favorable and unfavorable features.

2.1.   1980 PEP

The 1980 PEP was designed to study the net population undercount for each state and the 23 largest metropolitan areas. The PEP consisted of essentially two samples (termed P- and E-samples in what follows) and a matching process which used dual-system estimation to produce net undercount estimates. A detailed description of the PEP can be found in Cowan and Bettin (1982). The first sample consisted of about 186,000 persons in households in an ongoing monthly labor force survey in which a roster of persons in the households was obtained via a supplementary interview. The address was geographically coded to census geography. In fact, two separate, nonoverlapping monthly samples, April and August, were canvassed in this manner. However, no attempt has been made to combine the results, and each sample has been treated separately with respect to dual-system estimation. Each of these monthly samples are termed P-samples in the discussion that follows. The other sample consists of a sample of about 231,000 persons selected from the 1980 census from within the same selected primary sampling units associated with the P-sample and is termed the E-sample.

The PEP matched cases in the P-sample to the census files in the general location of the geocoded P-sample address. A status of matched or nonmatched was assigned to each person. Persons with a nonmatched status were sent back into the field for follow-up and then rematched to the census. All cases whose status (matched/not matched) could not be ascertained after the second match had a status imputed. Variation in the treatment of nonresponse cases and the manner of status imputation resulted in several different P-sample estimates.

The underlying concept of dual-system estimation is to conduct two independent listings of the population and to measure those that are observed in both listings. In our context, one listing of the population is accomplished by the census and the other is accomplished by the P-sample. However, direct use of the census counts in dual-system estimation is not feasible. The census operation includes in its count persons imputed on the basis of vague information and then allocates characteristics to them. Such persons could not be matched and were subtracted from census counts. In addition, an estimate of persons coded to incorrect geography, persons out of scope, and persons otherwise erroneously enumerated in the census was obtained via the E-sample and subtracted from the census count. In the E-sample procedure, interviewers returned to the census households. Persons not at the housing unit were followed up or neighbors were asked their whereabouts on census day. As in the P-sample, differing treatment of noninterviews and imputation of enumeration status resulted in several E-sample estimates.

Combinations of P- and E-sample treatments have resulted in 12 dual-system estimates of total population by age, race, and sex categories at the U.S. level and with lesser detail at the state and substate level. The particular combination of treatments used in our modeling efforts below is termed PEP 3-8 which is based on the April labor force survey sample. Our use of PEP 3-8 estimates (as opposed to any other PEP estimate) was mostly arbitrary. The PEP 3-8 procedure was the designated one prior to implementation of the PEP program. We used PEP 3-8 as an illustration although other PEP estimates are equally viable. In this P-sample all noninterviews are adjusted by a weighting procedure that assumes that the noninterviewed are similar to the interviewed. Also, match status of unresolved cases (those remaining after follow-up) were imputed using as a pool of donors those cases initially sent to follow-up and whose status subsequently were resolved. The E-sample cases lacking enumeration status after follow-up were given to the post office for resolution. Those cases not resolved were imputed using donor pools of like persons whose status were resolved by the post office.

For a particular category, let

$N_c \equiv$ census count of population
$N_p \equiv$ P-sample based estimate of population
$EE \equiv$ E-sample based estimate of census population erroneously enumerated
$M \equiv$ P-sample based estimate of population matched and
$II \equiv$ census count of population inputed

Then, the dual-system estimator of population total used in the PEP is $\hat{N}$ where

$$\hat{N} = N_p(N_c - EE - II)/M$$

and the net undercount is defined as $\hat{Y} = (\hat{N} - N_c)/\hat{N}$. When estimating for a particular geographic area, the categories used were age–race–sex within the area. Depending on the size of the area, the categories were collapsed until an adequate amount of sample cases were realized. Both P- and E-sample estimates include ratio adjustment.

According to Cowan and Bettin (1982), the proportion of cases in the sample that are missing data is larger than the estimated net undercount. For example, for PEP 3-8, the percentage of total persons, Black persons, non-Black Hispanic persons, and other persons requiring imputation were 4.1%, 7.2%, 7.3%, and 3.6%, respectively. The estimated net under-count for the same categories were 0.8%, 5.2%, 4.1%, and −0.1%. Consequently, the manner of imputation can have a major effect on the final estimates. There is some doubt as to whether independence is actually achieved in the PEP. Without independence the PEP estimates are biased. In addition, the listings are assumed to cover the entire population under consideration so as to yield a positive probability of response from every individual. It is questionable whether this was achieved in the PEP because the P-sample suffers from noncoverage. Despite these deficiencies, the PEP provides the only direct estimates of net undercount and gross errors at the sub-U.S. geographic level.

2.2. Demographic Analysis

The second data source for measuring undercount levels in the 1980 census is the method of demographic analysis. Demographic analysis provides national estimates of the population and of net undercount classified by age, sex, and race. As a tool for census evaluation, demographic analysis involves the combination of different types of demographic data to develop estimates for the population as of the census date, then the estimates are compared with the corresponding census counts. The particular procedure used to estimate the coverage for the various demographic subgroups depends primarily on the nature of the available data. For the population under age 45 in 1980, estimates of the resident population and coverage are based directly on birth, death, immigration, and emigration statistics and estimates. For the population aged 65 and over in 1980, estimates are developed from aggregate Medicare statistics adjusted for underenrollment in the Medicare files. For the population aged 45–64, the coverage estimates are based on population estimates derived primarily from the analysis of previous censuses. [See Passel and Robinson (1984) for discussion of the demographic method of estimating coverage.]

Since it has been estimated that at least 2 million undocumented aliens were counted in the 1980 census (Warren and Passel, 1983), an allowance for undocumented immigration must be added to the estimated resident population based on demographic analysis to obtain estimates of net undercount of the total population (legal and undocumented residents). The problem of undocumented immigration is a major source of uncertainty in the demographic estimates of coverage, especially for the non-Black population. For our purposes we assumed a level of 3.5 million illegal aliens assigned to age–race–sex categories on the basis of estimates derived by Warren and Passel (1983). The level of illegal aliens assumed here sets the demographic analysis estimate total population figure to approximately equal the PEP 3-8 total population estimate. From a small area estimation point of view, the lack of sub-U.S. undercount estimates and the illegal immigration problem are important drawbacks of demographic analysis.

2.3. 1980 Census

The 1980 Census provides much small area data in the way of population, housing, and administrative data that are possibly associated with undercount. In addition to age–race–sex counts at small geographic levels, urbanicity, labor force status, education, migration, language, income source, housing unit ownership, housing unit density, address list source, mail returns, substitution, and allocation counts of persons are examples of characteristics available for adjustment usage. Such data are tabulated to the district office level at present; the district office (DO) being the smallest level at which PEP 3-8 estimates are available. In the following section, we utilize the data at the DO level to model undercount and evaluate some of the adjustment methods. The DO is the administrative unit that was used to collect census information.

3. ADJUSTMENT METHODS

The adjustment methods considered to date are either of the synthetic or regression type. Variations of either type arise from the manner in which data resources are used. For example, net undercount adjustment factors at the total U.S. level could be used in a synthetic adjustment procedure by age–race–sex to provide substate level estimates of total population assuming a level of illegal immigration using demographic analysis. Synthetic adjustment could also be used by raking regional PEP 3-8 age–race–sex cell undercounts to state marginals and obtaining individual state age–race–sex cell adjustment factors for application to substate census data. Regression models using net undercount as the dependent variable could also be used to obtain substate estimates of total population adjusted for undercount.

3.1. Synthetic Estimation Using Demographic Analysis Estimates

Despite the limitations of demographic analysis data such as the unknown level of illegal immigration and the lack of sub-U.S. detail, synthetic estimation using demographic analysis estimates was investigated and compared with the 1980 census results at the state and DO level under the assumption that the corresponding PEP 3-8 estimates were the "truth." Some of these shortcomings could be reduced or eliminated in future census years. For example, the estimate of the segment of persons 45–64 years old would be reduced to the segment 55–64 years old in the next census and passage of a proposed bill in Congress could provide enough sanctions to enable a count of the number of illegals presently in the country and deter future illegal immigration. Finally, an advantage of synthetic estimation based on demographic analysis is that it could be done in a timely manner.

Basically, construction of demographic analysis-based synthetic estimates of total population for an area consists of two steps. In the first step, an adjustment factor for a given age–race–sex cell at the U.S. level is computed as a ratio of the demographic analysis figure to the census figure. In the second step, the corresponding census count of persons in the cell in the small area is multiplied by the relevant factor, and such products are summed over all cells in the areas. The assumption underlying this process is that undercount for the elements in the cell is uniform over all small areas. This assumption is questionable because it is likely that, for example, minorities in suburban areas are undercounted at a much lower rate than minorities in urban areas. Some comparisons have been made with the census with regard to total population and are presented below. We looked at the performance of three different synthetic estimators with regard to some measures proposed by Schirm and Preston (1984) as well as some other measures.

The three synthetic estimators labeled DA1, DA2, and DA3 differ in the manner of treatment of the Hispanic minority. In DA1, the Hispanics were combined with the non-Black group. In DA2, the Hispanics were treated separately by assuming they had the same adjustment factors as the Black group. In DA3, the PEP 3-8 undercount rate for Hispanics was used. In all three estimates the same Black adjustment factor was used. In DA2 and DA3, the non-Black, non-Hispanic group (termed other) factor was computed in a straightforward manner by removing the "corrected Hispanic count" from the non-Black demographic analysis figure. Table 1 displays the performance of DA1, DA2, DA3, and the census as estimates of total population for states with respect to the measures listed below and defined in the Appendix:

MARE ≡ mean absolute relative error

RSADP ≡ ratio of the sum of absolute differences of the adjustment proportions to the census proportions

PI ≡ proportion improvement after adjustments

RNAC ≡ ratio of the number of adjusted estimates within an interval of "truth" to the number of census counts within the same interval

RAC ≡ ratio of adjusted population estimates within an interval of "truth" to the census counts within the same interval

The measures in Table 1 favor DA2 over the other synthetic estimates as well as the census. Table 2 displays the same estimators and measures when estimation of total population of district offices is of interest. A minor change in the coverage is that the population under consideration is the noninstitutional population. The synthetic estimates do better than the census at this lower level but not nearly as well as at the state level. Note especially that the MARE has at least doubled for all methods.

The results in Table 2 assume that PEP 3-8 provides the correct population counts. In the absence of any other direct substate estimates of total population such comparisons are the best we can do.

Table 1. Comparison of Census and Adjustment Methods (States)

	Census	DA1	DA2	DA3
I. MARE[a]	0.0124	0.0119	0.0110	0.0112
II. RSADP[b]		1.014	1.142	1.088
III. PI[b]		0.505	0.707	0.688
IV. RNAC[b,c]		1.100 (22)	1.250 (25)	1.250 (25)
V. RAC[b]		1.181	1.113	1.113

[a]A smaller number is considered better.
[b]A larger number is considered better.
[c]Numbers in parentheses are counts of states falling in the interval.

Table 2. Comparison of Census and Adjustment Methods (DO)

	Census	DA1	DA2	DA3
I. MARE[a]	0.0328	0.0308	0.0300	0.0300
II. RSADP[b]		1.031	1.051	1.050
III. PI[b]		0.535	0.559	0.556
IV. RNAC[b,c]		1.078 (236)	1.123 (246)	1.123 (246)
V. RAC[b]		1.083	1.137	1.139

[a]A smaller number is considered better.
[b]A larger number is considered better.
[c]Numbers in parentheses are counts of DOs falling in the interval.

3.2. Regression Estimation Using PEP Data

Most of the modeling that follows is based on district office PEP 3-8 estimates of population. Several different regression models have been produced and are compared as to how well they predict district office population, assuming PEP 3-8 estimates are the "truth." While regression and synthetic estimation are considered separately here, it is important to point out that should regression be chosen as an adjustment method, synthetic estimation would also be playing a role in adjusting down to lower levels of aggregation. Two types of regression modeling are subsequently described. The first consists of several unweighted linear regressions and the second involves work by Ericksen and Kadane (1985) using a Bayesian hierarchical model.

Four hundred fourteen of the 422 district offices were used in all of the modeling work based on district offices. It was necessary to eliminate eight of the district offices due to insufficient sample size. Three models of net undercount using unweighted linear regression will be described below and compared later. The assumed model for the three equations is $Y = X\beta + \epsilon$, where $\epsilon \sim N(0, \sigma^2 I)$. The carrier variables, X, that predict percent net undercount, **Y**, are carrier variables formed from census tabulations. The carrier variables selected in the models that follow were chosen based on expert opinions as well as stepwise regresion procedures. All variables used are expressed in percent.

In the first two models, all 414 district offices were used to form both equations:

$$Y = -0.36 + 0.17(\text{MINRENT}), \qquad R^2 = 0.27, \quad S = 4.1 \qquad (1)$$

$$Y = 1.55 + 0.20(\text{MINRENT}) - 0.11(\text{NOHS}), \qquad R^2 = 0.29, \quad S = 4.0 \qquad (2)$$

where MINRENT = percentage of nonvacant renter-occupied housing that is minority

NOHS = percentage of total population that has not attended high school

Although model (2) does not appear to be significantly better than model (1), model (2) does seem to be a slightly better job predicting district office population as can be seen in Table 3.

While we would agree that it does not seem likely that the percentage minority renter variable alone explains the undercount problem fully, it does appear to be the only carrier variable we feel we can justify including from a model selection viewpoint. One of the ways we examined this issue was to generate dummy noise variables as suggested by

Table 3. Comparison of Adjustment Methods Using Unweighted Models[a]

	Model			
Measure	(1)	(2)	(3)	(4)
I. MARE[b]	0.0121	0.0115	0.0100	0.0100
II. RSADP[c]	1.481	1.626	1.524	1.515
III. PI[c]	0.607	0.644	0.688	0.550
IV. RNAC[c,d]	1.200 (24)	1.200 (24)	1.300 (26)	1.350 (27)
V. RAC[c]	1.040	1.117	0.978	1.001

[a]This table and table 4 are designed to make comparisons between possible adjustment models and should not be interpreted as a definitive statement that these models are better than the census.
[b]A smaller number is considered better.
[c]A larger number is considered better.
[d]Numbers in parentheses are counts of states falling in the interval.

Miller (1984). Then using the regression by the leaps and bounds procedure Furnival and Wilson (1974), we found the best 10 equations of two carrier variables based on the R^2 criterion. While model (2) was determined the best of the two carrier variable models with an R^2 of 0.29, the fifth best two carrier variable model had an R^2 of 0.27 with one of the carrier variables being one of the five dummy noise carrier variables. With noise doing almost as well as the percentage of the population not attending high school, we have further evidence of the large variability in the district office data. This is a major problem. Owing to the large variability it is difficult to fit models with reasonable carrier variables. While undercount is most likely a function of many different factors, given the district office data from 1980 we do not have evidence as to what those factors are except to say that there appears to be a relationship between undercount and minority renters at the district office level. Considering the results from the central city regression to be given, we may even conjecture that it is the central cities that are dictating this relationship.

In forming the third model, the 414 DOs were split into three groups each represented by its own model. The groups were chosen based on whether the district office was centralized, decentralized or conventional:*

$$\text{Net undercount (centralized)} = 19.16 + 0.18(\text{MINRENT})$$
$$- 0.26(\text{LISTCOR}), \quad R^2 = 0.38$$

$$\text{Net undercount (decentralized)} = -0.68 + 0.14(\text{CROWD})$$
$$+ 0.23(\text{BLMALE}), \quad R^2 = 0.04$$

$$\text{Net undercount (conventional)} = -2.98 + 0.11(\text{URBAN})$$
$$- 0.42(\text{CROWD})$$
$$+ 1.59(\text{FOR7580}), \quad R^2 = 0.51 \quad (3)$$

where MINRENT = percentage nonvacant renter-occupied housing that are minority
LISTCOR = percentage of occupied housing units that were listed correctly before census day

*Centralized DOs are located in large cities and canvassed by mail; conventional DOs are located in rural areas and canvassed via enumerators; decentralized DOs were canvassed by mail and constitute the bulk of the DOs.

CROWD = percentage of housing units with more than one person per room

BLMALE = percentage of population that are Black males aged 15–39

URBAN = percentage of total population that is urban

FOR7580 = percentage of total population foreign born and entering the United States between 1975 and 1980

As can be seen from equation (3), the minority renter variable, while important in the central city regression, does not appear in the decentralized or the conventional district office equations even though it is the variable most associated with undercount based on the combined set of district offices. While both the centralized and conventional areas can be modeled somewhat adequately, it was not possible to find an adequate model for the decentralized district offices. (We note that roughly two-thirds of their absolute net undercounts were less than 2%). While other groupings of the district offices based on different variables were attempted, the results were not as favorable.

The second method as advocated by Ericksen and Kadane involves the application of Bayesian hierarchical regression models for adjusting the census. These models were developed by Lindley and Smith (1972). Letting $Y = (Y_1, \ldots, Y_N)^T$ denote the vector of percentage net undercount estimates from the district offices, at the first level of the Bayesian hierarchical model it is assumed that

$$\mathbf{Y} \sim N(\boldsymbol{\theta}, \mathbf{D})$$

$$\boldsymbol{\theta}^T = (\theta_1, \ldots, \theta_N)$$

is a vector of mean values for $\mathbf{Y}$, and $\mathbf{D} = \mathrm{diag}(d_{11}, \ldots, d_{NN})$ is a diagonal matrix of the variances of the net percentage undercount estimates, which are assumed to be known. Although the true values of the d_{ii}'s are unknown, they have been taken to be equal to their survey estimates in Ericksen and Kadane's analysis. In addition to this approach, we have experimented with values of the d_{ii}'s obtained from empirical models.

At the second stage in the hierarchical model it is assumed that

$$\boldsymbol{\theta} \sim N(\mathbf{X}\boldsymbol{\beta}, \sigma^2 \mathbf{I})$$

where $\mathbf{X}$ is a matrix of p carrier variables, $\boldsymbol{\beta}$ is a vector of unknown parameters, and the value of σ^2 is assumed to be known. In their analysis, Ericksen and Kadane used percentage minority, percentage conventionally enumerated, and the crime rate as carrier variables to explain percentage net undercount for states and cities. In our research we are experimenting with alternative carrier variables in addition to those considered by Ericksen and Kadane. In their analysis as in ours, the true value of σ^2 is actually unknown but is taken to be equal to its maximum likelihood estimate.

At the final and third level of the Bayesian hierarchical model it is assumed that

$$\beta \sim N(\gamma, \Omega)$$

This stage is required to express knowledge about how the carrier information, X, explains the mean net undercount vector, θ. The matrix Ω^{-1} denotes how precise this knowledge is and, as in Ericksen and Kadane's analysis, we let $\Omega^{-1} = 0$ denoting that our knowledge is uninformative.

Using this Bayesian hierarchical formulation, the estimate of percentage net undercount is taken to be the posterior mean of θ:

$$[D^{-1} + \sigma^{-2}I]^{-1}[D^{-1}Y + \sigma^{-2}X\hat{\beta}]$$

That is, the Bayesian estimate of percentage net undercount is a mixture of the survey estimates, Y, and the modeled predictions, $X\hat{\beta}$, where $\hat{\beta}$ is a weighted least squares estimate.

Due to our interest in evaluating the carrier values chosen by Ericksen and Kadane at the district office level and comparing it to our other previously mentioned models, we fit variables very similar to theirs, the only change being that we substituted percentage migration for the crime variable. This was necessary because crime rates are not available at the district office level. While the model was to be fit to a state and central city data set that did in fact have the crime variable, our intention was to predict district office results. Percentage migration was selected as a reasonable proxy for crime:

$$Y = -2.58 + 0.08 \,(\text{MIN}) + 0.02 \,(\text{CONV}) + 0.04 \,(\text{MIGR}) \qquad (4)$$

where MIN = percentage of the total population that are Black or Hispanic
CONV = percentage of the area enumerated conventionally
MIGR = percentage of the population over 5 years old who did not live in the same house 5 years ago

The weighted model was estimated from the state and central city data using estimated rather than known variances. To investigate how models (1)–(4) compare when they are used to predict district office population counts, we used the same measures used in the synthetic estimation section. Again we treat estimated PEP 3-8 state population counts as "truth" comparing them to the district office predicted values summed to the state level.

As described previously the Ericksen and Kadane estimates are a mixture of the survey estimates and the modeled predictions. The modeled predictions are based on the linear model $Y = X\beta + \epsilon$, where $\epsilon \sim N(0, \sigma^2 I + D)$, D being a diagonal variance covariance matrix whose elements are the estimated variances of Y from the PEP. In the work below we will be comparing the following two models:

$$Y = 0.22 + 0.11(\text{MINRENT}) \tag{5}$$

$$Y = -1.90 + 0.06(\text{MIN}) + 0.003(\text{CONV}) + 0.04(\text{MIGR}) \tag{6}$$

Models (5a) and (6a) in Table 4 consists of predictions based on equations (5) and (6). Models (5b) and (6b) are mixtures of the direct estimates and their respective modeled predictions. According to Table 4 model (6b) appears to be the best although not by very much.

Since model (5b) using the minority renter variable does almost as well as the three variable model (both using estimated standard errors), it probably could be used without much if any loss in the precision of the adjustment results.

Table 4. Comparison of Adjustment Methods Using Both Weighted Models and Ericksen and Kadane Models (Based on 46 States)[a]

	Model			
	(5a)	(5b)	(6a)	(6b)
I. MARE[b]	0.0112	0.0092	0.0104	0.0088
II. RSADP[c]	35.707	39.420	35.174	36.828
III. PI[c]	0.758	0.758	0.758	0.758
IV. RNAC[c,d]	1.316 (25)	1.579 (30)	1.421 (27)	1.632 (31)
V. RAC[c]	1.430	1.460	1.396	1.468

[a]Because eight DOs did not have sufficient sample size to produce estimates of undercount, the states containing them had to be removed from this table.
[b]A smaller number is considered better.
[c]A larger number is considered better.
[d]Numbers in parentheses are counts of states falling in the interval.

As mentioned in our discussion of the Ericksen and Kadane work, an assumption of known variances is made. In our work, presented here, we have only estimated variances. We have, however, looked into the possibility of using modeled variances instead of the estimated variances but without much success.

3.3. Synthetic Estimation Using Post Enumeration Survey Data

Examples of synthetic estimates for undercount adjustment using a different PEP estimate than PEP 3-8 have been illustrated in Diffendal, Isaki, and Malec (1982) and Diffendal, Isaki, and Schultz (1984). In that application, regional PEP age–race–sex distributions were first raked to PEP state estimated marginals. The resulting cell estimates were used to construct state age–race–sex adjustment factors. It is of interest to repeat the process with the PEP 3-8 data and thereby construct DO estimates and compare them as in Table 2. We intend to do this in the future.

A somewhat related procedure to that previously described is due to Tukey (1981) and briefly reported by Ericksen and Kadane (1985). Rather than design a postenumeration survey (PES) to provide jurisdictional estimates of undercount, for example, regions, states, and large cities, the authors suggest designing a PES to provide estimates of undercount that satisfy the assumptions underlying synthetic estimation. This implies grouping together "areas" that are believed to have the same undercount rate. For example, some of the major variables correlated with undercount are minority and rural–suburban–central city. In this setting, the rural areas of South Dakota, North Dakota, Nebraska, and Iowa would be presumed to have similar undercount rates. The central cities of Baltimore, MD and Washington, DC could also be grouped together.

In this procedure, direct estimates of undercount for "areas" are divided by estimated census counts for the same "areas." These resulting adjustment factors are then regressed on related carrier variables, and the regression estimate is mixed with the adjustment factor to produce a final adjustment factor. The final adjustment factors are applied to census counts to obtain synthetic estimates. The suggested adjustment scheme contains both a synthetic estimation and a regression component. We use the word area in quotes to distinguish between traditional geographic areas and those likely to arise in the proposed methodology. It should be recognized that there is much similarity in the ideas presented here with those in the paper by Cohen and Kalsbeek (1974).

As a preliminary step in studying the above synthetic/regression procedure, as assume that the PES will again be a dual-system but that the sampling units at the last stage will be blocks. The issues of forming sampling strata as well as adjustment "areas" require study. We briefly describe a simulation procedure that we intend to pursue using 1980 census data that are felt to be associated with undercount. Some of the variables to be considered are allocations by race; urban, rural; race; female-headed households; renter occupied; all levels of geography; etc. For our study, some of the variables will be used as a pseudo-undercount variable, while others will be used to form sampling strata and adjustment "areas" and in the regression modeling of adjustment factors. It is necessary to create a pseudo-undercount variable for the study because direct estimates of census undercount at the "area" level are not likely to be available. Using the allocation variable plus the census count as a pseudovariable for the true number in the population, we can measure the error of the procedure in estimating total population at the enumeration district (ED) level, place level, county level, and so on. In this case we are assuming that the allocation variable has a distribution similar to that of the actual undercount and we are assessing the model error due to failure of the assumptions underlying the synthetic procedure. Since the adjustment factors will be estimated in practice (using a dual-system estimator in the numerator), a way must be found to simulate the sampling and model errors in the estimation of the adjustment factor. We are currently trying to solve this issue.

While the basic sampling unit is the block, without a special tabulation of the census variables to the block level, the data variables mentioned previously are available at the ED level (combination of from 1 to 20 blocks) only. We plan to conduct our preliminary study at the ED level initially, then proceed to a block level analysis.

4. SUMMARY

We have attempted to present a brief account of our current efforts in developing methods for census undercount adjustment in small areas. In the space allotted, it was not possible to present other issues that affect adjustment. For example, the adjustment method must allow for the timeliness of census operations and its operating schedule. The adjustment method must be able to produce output consistent with census publication output and be internally consistent as well. Methodology needs to be developed to add or delete persons for census files in accordance with undercount adjustment estimates. All of these are currently being addressed with regards to a test of adjustment-related operations to be conducted next year.

Finally, we are currently researching how to assess the effectiveness of each of our adjustment methodologies. Determining which methodology is the best will pose special difficulties since the standard by which we would like to rank our methodologies, the actual population sizes in specific geographical areas, is unknown. All of our efforts have essentially used variables with some deficiencies. It has been suggested that since the Black population is affected very little by illegal immigration that demographic analysis U.S. figures be used. Use of such a standard (if indeed it is true) may be satisfactory for the total Black population, but the problem of a standard for other races remains.

APPENDIX: DEFINITION OF MEASURES

1. $\text{MARE} = L^{-1} \sum_{i=1}^{L} |\text{PEP}_i^{-1}(E_i - \text{PEP}_i)|$

where $E_i =$ denotes the estimated total of area i using method E
$L =$ number of areas.

2. $\text{RSADP} = (\text{PSAE}^{\text{E}})^{-1}(\text{PSAE}^{\text{C}})$

where $\text{PSAE}^{\text{C}} = \sum_{i=1}^{L} |P_i^{\text{C}} - P_i^{\text{T}}|$

$\text{PSAE}^{\text{E}} = \sum_{i=1}^{L} |P_i^{\text{E}} - P_i^{\text{T}}|$

$P_i^{\text{C}} = \left(\sum_{i=1}^{L} \text{census}_i \right)^{-1} \text{census}_i$

$P_i^{\text{T}} = \left(\sum_{i=1}^{L} \text{PEP}_i \right)^{-1} \text{PEP}_i$

$P_i^{\text{E}} = \left(\sum_{i=1}^{L} E_i \right)^{-1} E_i$

3. $PI = \left(\sum_{i=1}^{L} \text{PEP}_i \right)^{-1} \sum_{i=1}^{L} \text{IMPV}_i$

where $\text{IMPV}_i = \begin{cases} \text{PEP}_i & \text{if } |P_i^{\text{E}} - P_i^{\text{T}}| < |P_i^{\text{C}} - P_i^{\text{T}}| \\ 0 & \text{otherwise} \end{cases}$

4. $\text{RNAC} = C^{-1}E$

where $E = \sum\limits_{i=1}^{L} R_i, \qquad C = \sum\limits_{i=1}^{L} S_i$

$$R_i = \begin{cases} 1 & \text{if } E_i \in D_i \\ 0 & \text{otherwise} \end{cases}$$

$$S_i = \begin{cases} 1 & \text{if census}_i \in D_i \\ 0 & \text{otherwise} \end{cases}$$

$V(\text{PEP}_i) = $ estimated variance of PEP_i

5. $\text{RAC} = (C')^{-1}E'$

where $E' = \sum\limits_{i=1}^{L} R'_i, \quad C' = \sum\limits_{i=1}^{L} S'_i$

$$R'_i = \begin{cases} \text{PEP}_i & \text{if } E_i \in D_i \\ 0 & \text{otherwise} \end{cases}$$

$$S'_i = \begin{cases} \text{PEP}_i & \text{if census}_i \in D_i \\ 0 & \text{otherwise} \end{cases}$$

REFERENCES

Bailar, B. A. (1983), Affidavit submitted to U.S. District Court, Southern District of New York, in Cuomo v. Baldridge, 80 Civ., 4550 (JES).

Bryce, H. J. (1980), "The Impact of the Undercount on State and Local Government Transfers," Conference on Census Undercount, U.S. Government Printing Office, Washington, DC, pp. 112–124.

Chandrasekar, C. and Deming, W. E. (1949), "On a Method of Estimating Birth and Death Rates and the Extent of Registration," *Journal of the American Statistical Association*, **44**, 101–115.

Coale, A. J., and Rives, N. W., Jr. (1973), "A Statistical Reconstruction of the Black Population of the United States, 1880–1970: Estimates of True Numbers by Age and Sex, Birth Rates, and Total Fertility," *Population Index*, **39**(1), 3–36.

Coale, A. J. and Zelnik, M. (1963), *New Estimates of Fertility and Population in the United States*, Princeton University Press, Princeton, NJ.

Cohen, S. B., and Kalsbeek, W. (1977), "An Alternative Strategy for Estimating the Parameters of Local Areas," *Proceedings of the Social Statistics Section of the American Statistical Association*, Washington, DC.

Cowan, C. D., and Bettin, P. J. (1982), "Estimates and Missing Data Problems in the Postenumeration Program," Technical report, U.S. Bureau of the Census, Washington, DC.

Diffendal, G. J., Isaki, C. T., and Malec, D. J. (1982), "Examples of Some Adjustment Methodologies Applied to the 1980 Census," Technical report, U.S. Bureau of the Census, Washington, DC.

Diffendal, G., Isaki, C., and Schultz, L. (1984), "Small Area Adjustment Methods for Census Undercount," invited papers to the Data Users Conference on Small Area Statistics, U.S. Department of Human Services, Washington, DC, pp. 52–56.

Ericksen, E. P. (1974), "A Regression Method for Estimating Population Changes of Local Areas," *Journal of the American Statistical Association*, **69**, 867–875.

Ericksen, E. P., and Kadane, J. B. (1985), "Estimating the Population in a Census Year-1980 and Beyond," *Journal of the American Statistical Association*, **80**, 98–109.

Fay, Robert E., III and Herriot, Roger A. (1979), "Estimates of Income for Small Places-An Application of James-Stein Procedures to Census Data," *Journal of the American Statistical Association*, 79, 269–277.

Furnival, G. M., and Wilson, R. W., Jr. (1974), "Regression by Leaps and Bounds," *Technometrics*, **16**(4), 499–511.

Hill, R. (1980), "The Synthetic Method: Its Feasibility for Deriving the Census Undercount for States and Local Areas," Conference on Census Undercount, U.S. Government Printing Office, Washington, DC, pp. 129–141.

Lindley, D. V., and Smith, A. F. M. (1972), "Bayes Estimates for the Linear Model," *Journal of the Royal Statistical Society*, Ser. B, **34**, 1–19.

Miller, A. J. (1984), "Selection of Subsets of Regression Variables," *Journal of the Royal Statistical Society*, Ser. A, **147**, 389–425.

Passel, J. S., and Robinson, J. G. (1984), "Revised Estimates of the Coverage of the Population in the 1980 Census Based on Demographic Analysis: A Report on Work in Progress," paper presented at the Meetings of the American Statistical Association.

Purcell, N., and Kish, L. (1979), "Estimation for Small Domains," *Biometrics*, **35**, 365–384.

Robinson, J. G., and Siegel, J. S. (1979), "Illustrative Assessment of the Impact of Census Underenumeration and Income Underreporting on Revenue Sharing Allocations at the Local Level," paper presented at the Meetings of the American Statistical Association.

Schirm, A. L., and Preston, S. H. (1984), "Census Undercount Adjustment and the Quality of Geographic Population Distributions," Technical report, University of Pennsylvania.

Siegel, J. S. (1968), "Completeness of Coverage of the Nonwhite Population in the 1960 Census and Current Estimates, and Some Implications," *Social Statistics and the City*, Joint Center for Urban Studies of the Massachusetts Institute of Technology and Harvard University.

Siegel, J. S., and Jones, C. D. (1980), "The Census Bureau Experience and Plans," Conference on Census Undercount, U.S. Government Printing Office, Washington, DC, pp. 15–24.

Slater, C. M. (1980), "The Impact of Census Undercoverage on Federal Programs," Conference on Census Undercount, U.S. Government Printing Office, Washington, DC, pp. 107–111.

Tukey, J. W. (1981), Discussion of "Issues in Adjusting the 1980 Census Undercount," by Barbara Bailar and Nathan Keyfitz, paper presented at the Annual Meeting of the American Statistical Association, Detroit, MI.

U.S. Bureau of the Census (1982), "Coverage of the National Population in the 1980 Census by Age, Race, Sex," *Current Population Reports*, *Ser. P-23, No. 115*, U.S. Government Printing Office, Washington, DC.

Warren, R. (1981), "Estimation of the Size of the Illegal Alien Population in the United States," Agenda Item B of the meeting of the American Statistical Association–Census Bureau Advisory Committee, November.

Warren, R., and Passel, J. S. (1983), "Estimates of Illegal Aliens from Mexico Counted in the 1980 United States Census," paper presented at the Meetings of the Population Association of America.

An Evaluation of Small Area Estimation Methods: The Case of Estimating the Number of Nonmarried Cohabiting Persons in Swedish Municipalities

S. Lundström
Statistics Sweden

ABSTRACT

The Swedish Population and Housing Census gives, every five years and for every municipality, the number of nonmarried cohabiting persons. Such information, however, is almost completely lacking for the years between the censuses, despite a strong demand. This paper presents, in that context, an evaluation of some alternative small area estimation techniques presented in the statistical literature. By means of Monte Carlo simulations, asymptotically design unbiased estimators (direct, poststratified, and generalized regression estimators) and model-dependent, design-biased estimators (synthetic, SPREE, and composite estimators) are compared with respect to the square root of the relative mean square error, the standard error, and the design bias.

1. INTRODUCTION

With five years intervals, the Swedish Population and Housing Census provides the quantity $N_{.iq}$, which denotes the number of not married ("not married" refers to unmarried, divorced, widows, and widowers) people in municipality q ($q = 1, \ldots, 277$) belonging to class i ($i = 1, 2$)

with respect to cohabitational status (cohabiting, not cohabiting). Despite a strong demand, such information for years between censuses is almost completely lacking.

It is true that the Register of the Total Population (RTP) provides the number of married persons, but since nearly 11% of the adults are cohabiting without being married the RTP information is of limited value. Moreover, the Survey on Living Conditions (SLC) provides a yearly national estimate, but the sample size is too small (3400 persons) to give an acceptable estimate for each municipality. The expected number of observations is less than 10 for about 70% of the 277 municipalities.

It is not feasible to include the characteristic of cohabitation in the RTP, and it is too costly to increase the sample size in the SLC to yield the statistics we seek. Therefore, we try to develop estimators that combine data from different sources. In the literature we find several examples on rather successful attempts to use such model-dependent estimators as the synthetic estimator and the SPREE estimator.

Most small area estimators in use introduce a third classification of the population into H mutually exclusive and exhaustive classes, which can be based on age, sex, income, etc; they are labeled $h = 1, \ldots, H$. The population will thereby, in our case, be completely cross-classified into $277 \times 2 \times H$ cells with unknown cell sizes N_{hiq}, representing the number of not married people in municipality q with cohabitational status i, belonging to sex- and age-group h. The statistical problem is to estimate the quantities $N_{.iq} = \Sigma_h N_{hiq}$.

The computerized RTP register each month provides current information on the number of not married people in municipality q belonging to sex- and age-group h, $N_{h.q}$ ($= \Sigma_i N_{hiq}$). Moreover, we have current sample information from the SLC, which can be used to form estimates of $N_{hi.}$ ($= \Sigma_q N_{hiq}$). At the intercensal period we also known N'_{hiq}, which denotes the number of not married people in cell hig at the latest census.

The available data are rather extensive, and conditions are thus favorable for producing small area statistics.

In order to measure both the sampling error and the bias, Monte Carlo simulations are carried out. [Some of the findings of this paper have been published in Lundström (1984).]

2. DESCRIPTION OF THE MONTE CARLO SIMULATIONS

2.1. The Design of the Study

The Monte Carlo simulations are designed with the purpose of studying estimators for the totals $N_{.iq}$ at the 1980 census period. Thus, we know the parameter values and can for each sample repetition compare the estimates with the parameters and compute different quality measures.

The latest complete data, N'_{hiq}, will be retrieved from the 1975 census. This implies that the time between the computation of N'_{hiq} and the estimates $\hat{N}_{.iq}$ is at most five years. The current information $N_{h.q}$ is in the study taken from the 1980 census–not from the RTP. However, this will not affect the results because the difference between the RTP and the census is negligible with respect to this quantity.

The sample information in the simulation study is obtained in the following way: The interval $(0, 1)$ is divided into parts where the part corresponding to the cell hiq has the length N_{hiq}/N . Then n random numbers are drawn from a variable uniformly distributed on $(0, 1)$. Each random number is located in one and only one cell and n_{hiq} is obtained.

The SLC is based on a stratified sample with systematic sampling within strata. The population is divided into only two strata according to age. Among older people a larger fraction is drawn than among younger people. Thus the sampling design of the simulation study is different from that of the SLC, but this will not to any large extent affect the evaluation.

There are some differences between the SLC and the census in the definition of study variable, reference period, and data collection methods, which lead to a deviation between the expected SLC value and the census value. These differences are not reflected in the simulation study.

2.2. Estimators under Study

In a first part of the study six estimators are examined. Three of them are approximately design unbiased (ADU), while the others are model dependent and hence design biased. In a second part we also examine a composite estimator, which is a weighted function of an ADU estimator and a model-dependent estimator.

It should be observed that the estimators to be studied utilize different amounts of auxiliary information. Since an estimator that incorporates strong auxiliary information is expected to outperform an estimator that only incorporates weak auxiliary information (or none at all), some of the estimators under study could *a priori* have been excluded. However, we think it is of interest to measure the effect of different types and amounts of information.

(i) If neither the information N'_{hiq} nor the associated variable (age/ sex) is used, the Horvitz–Thompson (HT) estimator is an obvious (but poor) candidate:

$$\text{HT} = \frac{N_{...}}{n_{...}} \, n_{.iq} \tag{1}$$

(ii) By making use of the known quantities $N_{h.q}$, the following post-stratified (PST) estimator is close at hand:

$$\text{PST} = \sum_h \frac{N_{h.q}}{n_{h.q}} \, n_{hiq} \tag{2}$$

Obviously, this estimator is not defined for every possible sample, namely, when $n_{h.q} = 0$. One solution often recommended in the literature is to merge two or more poststrata. In the present simulation study, however, we have adopted a computationally simpler method: if, for some h, a zero count $(n_{h.q} = 0)$ is realized, the corresponding term is dropped when summing over h. This rule makes the PST estimator biased, especially for small sample sizes.

(iii) In a generalized regression approach Särndal (1981) has developed an ADU estimator:

$$\text{DM} = \sum_h \left\{ \frac{N_{h.q}}{n_{h..}} \, n_{hi.} + \frac{N_{...}}{n_{...}} \left(n_{hiq} - n_{h.q} \frac{n_{hi.}}{n_{h..}} \right) \right\} \tag{3}$$

(The estimator takes both the design and the model into consideration, therefore the abbreviation DM.)

This estimator is in effect the synthetic estimator (see SYNT, below) corrected for design bias (SYNT minus estimated design bias), and it can also be written

$$\text{DM} = \sum_h \left\{ \frac{N_{h.q}}{n_{h..}} n_{hi.} + \hat{N}_{h.q} \left(\frac{n_{hiq}}{n_{h.q}} - \frac{n_{hi.}}{n_{h..}} \right) \right\} \tag{4}$$

where

$$\hat{N}_{h.q} = \frac{N_{...}}{n_{...}} \, n_{h.q}$$

Remark 1. In (4) if we exchange $-\hat{N}_{h.q}$ for $N_{h.q}$, the DM estimator is transformed into the PST estimator. Hence, the larger the sample size, the smaller the difference between the two estimators.

Remark 2. With small sample sizes the DM estimator can give negative estimates. Such estimates would hardly be accepted in a real-life application, since they are not in the parameter space. (If negative estimates are replaced by, for example, zero, the DM estimator will be biased but have smaller variance.) In the present simulation study, however, negative estimates are accepted.

The preceding three estimators are all asymptotically design unbiased. We will now turn to some estimators that lack this appealing large sample property. However, they have other merits, which make them strong competitors. These latter estimators can all be considered as special cases of a class of estimators (structure preserving estimates–SPREE) proposed by Purcell (1979) in a categorical data analysis approach. One feature of this approach is the implicit assumption of a superpopulation model for the behavior of small domain frequencies over time.

In short, a SPREE estimator is defined through an adjustment of data from a previous point of time to given current marginal totals, while at the same time as far as possible preserving the structure of interaction between the variables as established at the previous point of time. Special cases will be spelled out in more detail below.

(iv) When at least two current margins are known, the SPREE estimators are the result of an iterative proportional fitting (IPF) procedure. In the present context this means that we have access to the current margins $\hat{N}_{hi.}$ $[= (N_{h..}/n_{h..})n_{hi.}]$ and $N_{h.q}$ (as well as to the complete data set N'_{hiq} from the previous point of time). In this situation the IPF algorithm can be written as follows.

The known previous data N'_{hiq} are taken as initial proxies, that is,

$$\hat{N}^{(0)}_{hiq} = N'_{hiq} \tag{5}$$

At the kth iteration we compute

$$_1\hat{N}^{(k)}_{hiq} = \frac{\hat{N}^{(k-1)}_{hiq}}{\hat{N}^{(k-1)}_{hi.}} \hat{N}_{hi.}$$

where

$$\hat{N}^{(k-1)}_{hi.} = \sum_q \hat{N}^{(k-1)}_{hiq}$$

and

$$\hat{N}^{(k)}_{hiq} = \frac{_1\hat{N}^{(k)}_{hiq}}{_1\hat{N}^{(k)}_{h.q}} N_{h.q}$$

where

$$_1\hat{N}_{h.q}^{(k)} = \sum_i {}_1\hat{N}_{hiq}^{(k)} \tag{6}$$

The iterative process is continued until some convergence criterion is satisfied (assume that this will happen when $k = k_0$); finally the SPREE (SPR) estimate is calculated as

$$\text{SPR} = \sum_h \hat{N}_{hiq}^{(k_0)} \tag{7}$$

(v) If the complete previous data, N'_{hiq}, and the current margin $\hat{N}_{hi.}$ but not the $N_{h.q}$ are known, we obtain the SPREE estimator

$$\text{SYNT1} = \sum_h \frac{N'_{hiq}}{N'_{hi.}} \hat{N}_{hi.}$$

where

$$\hat{N}_{hi.} = \frac{N_{h..}}{n_{h..}} n_{hi.} \tag{8}$$

This estimator will result from an IPF procedure where only step 1 in each cycle is performed. It can also be derived directly as a solution of the minimizing problem associated with SPREE estimation.

(vi) If only the current information $\hat{N}_{hi.}$ and $N_{h.q}$ are known, the SPREE estimator is given by

$$\text{SYNT} = \sum_h \frac{N_{h.q}}{N_{h..}} \hat{N}_{hi.} \tag{9}$$

This estimator is the result of the IPF procedure when all initial proxies have the same value.

2.3. Results

The SPREE estimators suffer from sampling error and design bias and, therefore, the quality measure must include these two quantities. The quality measure we have chosen to estimate in the simulation study is the square root of the relative mean square error (rel-MSE)

$$(\text{rel-MSE})^{1/2} = \left\{ E\left(100 \frac{\hat{N}_{.iq} - N_{.iq}}{N_{.iq}} \right)^2 \right\}^{1/2} \tag{10}$$

We also estimate the relative standard error (rel-se)

$$\text{rel-se} = \left\{ E\left(100\ \frac{\hat{N}_{.iq} - E(\hat{N}_{.iq})}{N_{.iq}} \right)^2 \right\}^{1/2} \tag{11}$$

and the relative bias (rel-bias)

$$\text{rel-bias} = E\left(100\ \frac{\hat{N}_{.iq} - N_{.iq}}{N_{.iq}} \right) \tag{12}$$

These quantities are associated in the following way

$$\text{rel-MSE} = (\text{rel-se})^2 + (\text{rel-bias})^2 \tag{13}$$

The variable of interest consists of only two categories. Since we know the sum $N_{..q}\ (= N_{.1q} + N_{.2q})$, we only present the results for the estimation of $N_{.1q}$, the number of nonmarried cohabiting people in each municipality.

The budget does not allow the inclusion of all 277 municipalities. Instead we examine several minipopulations. The largest of these consists of 55 municipalities. Our two Monte Carlo simulations consistent of (a) the selection of 400 samples, each of size $n \ldots = 1000$; (b) the selection of 200 samples of size $n_{...} = 5000$.

The results expressed in terms of the square root of rel-MSE are presented in Figures 1a and 1b where the minipopulation consists of the 18 largest municipalities; simulations for other minipopulations show a similar quality picture.

The figures show that the ADU estimators have a larger $(\text{rel-MSE})^{1/2}$ than the model-dependent estimators, when the sample size is 1000. Only the usual synthetic estimator, SYNT, has for some municipalities a larger $(\text{rel-MSE})^{1/2}$ than the ADU estimators.

When comparing the biased estimators, the importance of the auxiliary information appears quite clearly. If the previous information N'_{hiq} is lacking (i.e., the SYNT estimator), the problem with a large bias is obvious. The SPR estimator uses the current data, $N_{h.q}$, in contrast to the SYNT1 estimator, but the figures (and also Table 1) show that $N_{h.q}$ has no clear positive effect on the results.

The annual survey SLC mentioned earlier in this paper has a sample that contains about 1000 not married persons from the 18 largest municipalities. If we want estimates for these municipalities, and if we believe that $(\text{rel-MSE})^{1/2}$ is a relevant measure of the quality, then Figure 1a indicates that we ought to choose SYNT1 or SPR.

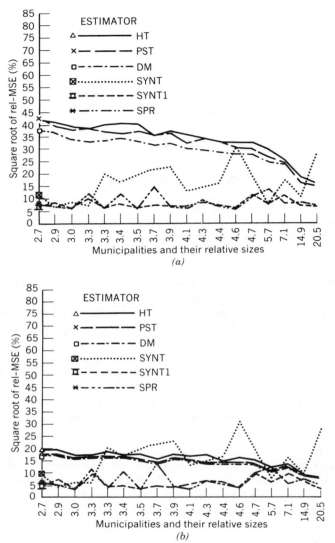

Figure 1. Square root of rel-MSE. Minipopulation: the 18 largest municipalities. (*a*) Sample size 1000, (*b*) sample size 5000.

Table 1. Average Root of rel-MSE for Different Minipopulations when the Sample Size is 1000

Minipopulation	Estimator					
	HT	PST	DM	SYNT	SYNT1	SPR
The 18 largest	33.9	32.2	29.3	16.3	7.8	8.5
"Sample 1"	39.8	37.1	34.9	21.2	13.2	11.3
"Sample 2"	33.7	30.9	28.4	12.4	11.5	11.7
The 55 largest	57.1	51.9	50.1	14.3	9.3	9.6

When the sample size is increased, the ADU estimators and the biased estimators are brought closer to each other. At the sample size 1000, the DM estimator has a smaller $(\text{rel-MSE})^{1/2}$ than PST, but at the sample size 5000 the difference is negligible.

The biased estimators change just slightly when we increase the sample size from 1000 to 5000, which means that the bias is the dominating error even in the case with the smaller sample size. Also, simulations have been carried out for two different samples of 18 municipalities (denoted "Sample 1" and "Sample 2") and for the 55 largest municipalities. In Table 1 we present the average $(\text{rel-MSE})^{1/2}$ when the sample size is 1000.

Table 1 shows that if you are looking for an approximately design unbiased estimator, the DM estimator will be a good choice. If your only demand is to have an estimator with a small average mean square error and you only have the current information $\hat{N}_{hi.}$ and $N_{h.q}$, then you should choose SYNT.

If you also have access to previous data, N'_{hiq}, the choice is less straightforward. One advantage with SPR compared to SYNT1 is that SPR directly provides estimates that add up to the current marginal totals $N_{h.q}$. On the other hand, SPR is more complicated to calculate than SYNT1.

The size of the ratio between the bias and the standard error is decisive for the possibility of computing confidence intervals. In Table 2 below this ratio for the biased (the relative bias for the HT and DM estimators is never important) estimators is displayed.

Table 2 shows that PST has a bias which, in general, cannot be neglected when confidence intervals are computed. The bias is caused by the rule used in the present simulation study when $n_{h.q} = 0$. However, the model-dependent estimators are much more affected by the bias, and, hence, a conventionally computed confidence interval will be quite misleading as a quality measure.

Table 2. Ratio between the Absolute Value of the Bias and the Standard Error for the Biased Estimators when the Sample Size is 1000 (Minipopulation: "Sample 1")

Municipality	Estimator				Relative Size of the Municipality (%)
	PST	SYNT	SYNT1	SPREE	
1	0.32	4.09	1.70	1.27	0.6
2	0.13	6.93	2.55	0.52	1.0
3	0.29	0.27	6.60	2.69	1.0
4	0.30	1.44	4.82	4.08	1.4
5	0.39	5.36	1.09	0.88	1.6
6	0.19	2.16	0.88	0.48	1.9
7	0.17	4.90	2.93	1.94	2.7
8	0.09	0.32	1.54	1.89	2.8
9	0.01	1.42	3.63	4.25	3.1
10	0.04	5.57	0.81	0.28	3.8
11	0.01	0.90	2.08	0.72	4.6
12	0.07	6.10	3.54	2.96	5.1
13	0.00	4.64	0.91	1.66	6.9
14	0.06	4.88	1.61	0.41	8.1
15	0.02	1.14	0.42	0.74	9.3
16	0.01	0.83	0.58	0.60	11.5
17	0.00	2.67	0.34	0.79	15.5
18	0.01	1.29	0.89	1.66	19.0

3. MONTE CARLO SIMULATIONS FOR COMPOSITE ESTIMATORS

The ADU estimators are approximately design unbiased but suffer from large sampling errors, while the model-dependent estimators are design biased with small sampling errors. In the literature there are several examples of attempts to construct composite estimators aiming to combine the strengths of each group of estimators [e.g., Schaible, Brock, and Schnack (1977) and Drew, Singh, and Choudhry (1982)]. They have in some cases performed well.

The estimators SYNT1 and SPR have a smaller relative mean square error than the other estimators previously examined. Therefore, it seems reasonable to choose one of those two. However, owing to wider general applicability, we will concentrate on estimators based on current data only. We have chosen the poststratified estimator, PST, and the synthetic estimator, SYNT.

One simple type of composite estimator can be written

$$\hat{N}_{.iq} = C_q \text{PST} + (1-C_q)\text{SYNT} \tag{14}$$

where C_q is an *a priori* fixed weight.

It is easily shown that the weight that minimizes the mean square error of (14) is

$$C_q^* = \frac{E\{(\text{SYNT-}N_{.iq})(\text{SYNT-PST})\}}{E(\text{PST-SYNT})^2} \tag{15}$$

which may be rewritten as

$$C_q^* = \frac{\text{MSE(SYNT)} - E\{(\text{SYNT-}N_{.iq})(\text{PST-}N_{.iq})\}}{\text{MSE(PST)} + \text{MSE(SYNT)} - 2E\{(\text{SYNT-}N_{.iq})(\text{PST-}N_{.iq})\}} \tag{16}$$

Obviously, the relevant quantities in expression (16) are not easily assessed; this is especially so for the expected cross-product. However, if this latter quantity is negligible relative to MSE(SYNT), the optimal weight simplifies [we mainly follow suggestions in Schaible (1979)] to

$$C_q^{**} = \frac{1}{1 + \text{MSE(PST)}/\text{MSE(SYNT)}} \tag{17}$$

The problem of determining the optimal weight is reduced to the assessment of the ratio MSE(PST)/MSE(SYNT).

Another type of composite estimates is

$$\hat{N}_{.iq} = \tilde{C}_q \text{PST} + (1 - \tilde{C}_q)\text{SYNT} \tag{18}$$

where $\tilde{C}_q$ is a sample-dependent weight. In an effort to arrive at a simple weight we have used the following crude lines of argument: Suppose $\text{MSE(PST}\,|\,n_{..q}) \doteq a/n_{..q}$, where a is a constant. Furthermore, suppose that we have $\text{MSE(SYNT)}|n_{..q}) \doteq b$, where b is a constant. If so, we might—with (17) in mind—try the weight

$$\tilde{C}_q = \frac{1}{1 + (a/bn_{..q})} = \frac{n_{..q}}{n_{..q} + a/b} \tag{19}$$

If a/b can be approximated from, for example, simulation studies on

Table 3. Weights in the Composite Estimators and Square Root of rel-MSE (Minipopulation: "Sample 3")

Munici-pality	Optimum Weight (a)	Optimum Weight in the 1975 Census (b)	$\dfrac{1}{1+105/E(n_{..q})}$ (c)	(rel-MSE)$^{1/2}$ for the Following Estimators: Composite (a)	Composite (b)	Composite (c)	PST	SYNT	Relative Size of the Municipality (%)
1	−0.04	0.07	0.05	9.2	11.9	11.0	75.4	9.5	0.6
2	−0.09	−0.02	0.05	14.3	14.8	16.6	74.4	15.2	0.6
3	−0.05	0.03	0.05	14.3	15.5	16.2	70.9	14.8	0.7
4	−0.03	−0.03	0.07	24.2	24.2	24.5	66.8	24.2	0.8
5	−0.06	0.07	0.07	13.5	15.7	15.5	67.0	14.0	0.8
6	−0.02	0.10	0.07	28.5	28.8	28.4	58.5	28.5	0.8
7	0.04	−0.05	0.07	27.7	27.6	27.7	61.4	27.6	0.8
8	0.07	−0.02	0.07	27.2	27.4	27.0	59.3	27.3	0.8
9	−0.03	−0.03	0.08	24.7	24.7	25.0	61.2	24.7	0.9
10	−0.02	−0.01	0.08	20.5	20.5	21.2	62.8	20.6	0.9
11	−0.03	−0.01	0.09	21.2	21.1	21.3	57.6	21.1	1.1
12	0.29	0.30	0.10	35.3	35.3	36.7	48.2	38.5	1.1
13	0.00	0.15	0.14	14.0	15.5	15.2	52.9	14.0	1.7
14	0.00	0.01	0.16	10.4	10.4	12.5	45.7	10.4	2.0
15	0.03	0.09	0.18	12.4	12.6	14.0	46.7	12.4	2.2
16	0.20	0.27	0.29	14.7	14.9	15.1	31.2	16.1	4.4
17	0.72	0.79	0.41	16.0	16.3	19.2	19.5	30.4	7.4
18	0.88	0.89	0.87	78.5	8.5	8.5	8.7	14.7	72.4

populations similar to the one under study, we have found a sample-dependent weight.

The simulation study is designed in the same way as the study previously reported. However, we have only carried out one Monte Carlo simulation consisting of a selection of 400 samples, each of size $n_{...} = 1000$. The minipopulation consists of a random sample of 18 municipalities that will be denoted "Sample 3" in the following.

In the simulation study we examine the composite estimator $\hat{N}_{.iq}$, using three different sets of weights:

(a) $C_q^* =$ optimal values for the current population

(b) $C_q =$ optimal values at the 1975 census;

(c) $\tilde{C}_q = 1/(105/n_{..}q + 1)$

where the value 105 ($= a/b$) is determined from the simulation study for the minipopulation "Sample 1."

In the first step the optimal C_q^* values in (a) are estimated by unbiased estimation of the numerator and denominator of (15) by using the 400 estimates. The same procedure is then repeated for the minipopulation from the 1975 census in order to estimate the C_q values in (b). In the second step (rel-MSE)$^{1/2}$ for the composite estimators are estimated in the same way as for the estimators reported previously. The $\tilde{C}_q$ values in alternative (c) are calculated for each sample. The results from the Monte Carlo study are presented in Table 3. Since the weight in alternative (c) varies from one sample to another we have calculated the quantity $1/\{1 + 105/E(n_{..q})\}$ in order to give the reader some idea of the size of the weights actually used.

We note that the synthetic estimator has a smaller (rel-MSE)$^{1/2}$ than the poststratified estimator in all of the municipalities but the two largest. All three composite estimators also provide estimates that, referring to (rel-MSE)$^{1/2}$, to a large degree resemble that of the synthetic estimator in all municipalities but the two largest, where the composite estimator is superior.

The composite estimator based on (b)-weights is for most municipalities a better estimator than the estimator based on (c)-weights. However, note that the use of alternative (c) provides a smaller (rel-MSE)$^{1/2}$ than both PST and SYNT for six municipalities.

Looking at the weights and the (rel-MSE)$^{1/2}$ we find that the composite estimator is rather insensitive to deviations from the optimum weight.

4. CONCLUSIONS AND PLANS FOR FUTURE WORK

The results of the Monte Carlo simulations for small area estimation of the number of nonmarried cohabiting persons in municipalities have led to the following conclusions:

(a) The model-dependent estimators are superior to the ADU estimators for "common" sample sizes with respect to $(\text{rel-MSE})^{1/2}$.

(b) Among the model-dependent estimators, SYNT1 and SPR show good potential.

(c) The DM estimator seems to be superior to the alternative ADU estimators included in the study.

(d) When only current information is available, the composite estimator seems to be a good choice.

Recently, a question on cohabitational status has been included in the Swedish Labour Force Survey (LFS). The LFS is based on a much larger sample than the SLC, and, therefore, the conditions are more favorable for the ADU estimators. On the other hand, the larger sample size can also be used to reduce the mean square error of the model-dependent estimators. Ongoing work, not reported here, shows that to a large extent it is possible to reduce the bias of the model-dependent estimates by clustering the municipalities in homogeneous groups according to previous data. Estimates can then be calculated for the municipalities in each group. This will increase the standard error (you use only the part of the sample belonging to the group of municipalities), but this increase will–at least in some groups–be much smaller than the reduction of the bias. We are at present working on the problem of finding a suitable clustering strategy. Finally, we intend to calculate and publish the small area estimates.

REFERENCES

Drew, J. D., Singh, M. P., and Choudhry, G. H. (1982), "Evaluation of Small Area Estimation Techniques for the Canadian Labour Force Survey," *1982 Proceedings of the Section on Survey Research Methods*, American Statistical Association, pp. 545–550.

Gonzales, M. E. (1973), "Use and Evaluation of Synthetic Estimates," *1973 Proceedings of the Social Statistics Section*, American Statistical Association, pp. 33–36.

Lundström, S. (1984), "Estimation for Small Domains: Two Studies Using Combined Data from Censuses, Surveys and Registers," *Statistical Review* **2**, 119–126 (Statistics Sweden).

Purcell, N. J. (1979), "Efficient Small Domain Estimation: A Categorical Data Analysis Approach," unpublished Ph.D. thesis, University of Michigan, Ann Arbor, MI.

Purcell, N. J., and Kish, L. (1979), "Estimation for Small Domains," *Biometrics*, **35**, 365–384.

Särndal, C. E. (1981), "Framework for Inference in Survey Sampling with Applications to Small Area Estimation and Adjustment for Nonresponse," *Bulletin of the International Statistical Institute*, **49**(1), 494–513.

Särndal, C. E. (1984), "Design-Consistent versus Model-Dependent Estimation for Small Domains," *Journal of the American Statistical Association*, **79**, 624–631.

Schaible, W. L. (1979), "A Composite Estimator for Small Area Statistics. Synthetic Estimates for Small Areas," *Synthetic Estimates for Small Areas*, NIDA Research Monograph 24, pp. 36–53.

Schaible, W. L., Brock, D. B., and Schnack, G. A. (1977), "An Empirical Comparison of the Simple Inflation, Synthetic and Composite Estimators for Small Area Statistics," *1977 Proceedings of the Social Statistics Section*, American Statistical Association, pp. 1017–1021.

PART 5

Panel Discussion

Discussion

T. Dalenius
Brown University

"Small is beautiful"

This note summarizes and elaborates on the comments I made during the panel discussion on May 23; it reflects to some extent comments by the other members of the panel and by the audience.

1. THE METHODOLOGICAL SETTING

1.1. The Notion of Small Area Statistics

Small area statistics, SAS for short, is commonly conceived of as statistics about such units as individuals, households, and farms, associated with the small (administrative) areas into which a larger (administrative) area has been divided. As an example, the small areas may be counties and the large area a state.

There is no reason why the scope of SAS should be restricted to *administrative* areas. It appears natural to extend the scope to encompass any kind of area, including, for example, the "drainage area" of a river system. It may even be argued that a further extension is called for, approximately corresponding to the terms "domains of study" and "breakdowns."

1.2. The Objectives

Consider a large area, divided into A small areas. In the interest of perspicuity, the statistics of interest are assumed to be "totals." Thus, for

257

the large area, there is a total Y, which may be a "count" (such as the total number of households) or a "magnitude" (such as the total household income). This total is the sum of A subtotals:

$$Y = Y_1 + Y_2 + \cdots + Y_a + \cdots + Y_A .$$

These subtotals—which will be referred to as "Y_a-statistics"—are either unknown, or known only with unacceptably low accuracy. The objectives of SAS are to provide reasonably accurate estimates of the Y_a-statistics.

2. ESTIMATING THE Y_a-STATISTICS

2.1. Two References

Realizing that "publication means significance," survey statisticians have in recent years devoted considerable efforts to develop a statistically sound methodology for SAS. Two pertinent references are: (1) Steinberg J. (ed.) (1979), *Synthetic Estimates for Small Areas. Statistical Workshop Papers and Discussion*. NIDA Research Monograph 24, National Institute on Drug Abuse, Rockville, MD, U.S. Government Printing Office, Washington, DC. (2) *Panel on Small-Area Estimates of Population and Income. Estimating Population and Income of Small Areas*, Committee on *National Statistics* (1980), National Academy Press, Washington, DC. This symposium is an additional demonstration of the responsibility that the survey statisticians show.

2.2. The Challenge

It is obviously possible to carry out A sample surveys in order to estimate the Y_a-statistics. This *direct* approach would in practice prove to be unsatisfactory: it may be prohibitively expensive (especially for large A) and/or too time consuming.

Instead, an *indirect* approach is used: the Y_a-statistics are estimated using somehow already available statistics for the small areas. In Section 2.3, an overview will be presented of some possible estimation techniques.

2.3. The Indirect Approach

The already available statistics is a set of X_a-statistics:

$$X_1, X_2, \ldots, X_a, \ldots, X_A$$

and hence the total X. These statistics may be the results of a census (not necessarily flawless), or they may be sample estimates.

I will consider three estimation techniques; they do not exhaust the options available. The three techniques have in common that they exploit an "estimation algorithm" $f(X_a)$, which reflects a (real or assumed) relationship between the Y- and X-statistics.

(a) Technique No. 1

An estimate of Y_a, $a = 1, \ldots, A$ is computed as

$$Y_a^* = f(X_a)$$

Depending on the structure of $f(\cdot)$ the estimation may be referred to as "regression estimation," "synthetic estimation," etc.

(b) Technique No. 2

If the total Y is known,

$$\tilde{Y}_a = Y_a^* \cdot \frac{Y}{\sum_a Y_a^*}$$

may be a better estimate than Y_a^*. The same may hold true if, instead of Y, an estimate Y^* is used.

(c) Technique No. 3

This technique uses—in addition to the X_a-statistics—y_a-data available from a sample survey. The volume of such data is "small"; hence, estimating the Y_a-statistics solely on the basis of the y_a-data is out of the question.

The y_a-data may, however, be exploited as follows. Estimates $\hat{Y}_a$ are computed using the y_a-data. These estimates are then combined with the estimates Y_a^* to yield the new estimates:

$$\tilde{Y}_a = wY_a^* + (1 - w)\hat{Y}_a$$

where $0 < w < 1$. In most applications, w would be close to 1, reflecting the fact that the y_a-data are sparse.

In what follows, the discussion will focus on Y_a^*.

2.4. The Choice of Estimation Algorithm

The three estimation techniques discussed in Section 2.3 may be characterized as "model-dependent." If $f(\cdot)$ faithfully reflects the relation between the Y_a- and X_a-statistics, these techniques may perform well. But if this condition is not fulfilled, the resulting estimates may be seriously deficient.

This raises the question: What is a proper basis for choosing a specific $f(\cdot)$?; or alternatively, what is it we should try to model?

It may seem near at hand to choose $f(\cdot)$ in a way which may be described as "data modeling": given a set of y_a-data, a function $f(\cdot)$ is chosen that provides a good fit to the y_a- and X_a-data.

Data modeling is sometimes justified by saying that "experience shows that data modeling works well." I think this is too shaky a ground for the choice of $f(\cdot)$. In support of this view, I will briefly consider a *general* illustration (i.e., one not in the realm of SAS), where "data modeling" breaks down.

Assume we have observed some phenomenon/property at two points of time. At time t_1, we have made the observation z_1 and at time t_2, the observation z_2, as shown in Figure 1. We want to make a prediction of the z-value z_3 at time t_3. No statistically meaningful prediction can be made *solely* on the basis of the data at hand: depending on the kind of phenomenon/property being observed, we may expect $z_3 > z_2$, $z_1 < z_3 < z_2$, or $z_3 < z_1$. What is clearly needed is a good understanding of the *mechanism*, which generated z_1 and z_2; this is tantamount to saying that we must have the specific subject matter knowledge. In order to enhance the prior knowledge we may already have, it should prove helpful to make use of techniques for (exploratory) data analysis, which have been developed in the last 10–20 years.

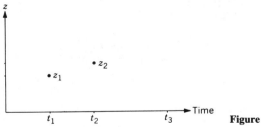

Figure 1.

3. THREE SUPPLEMENTARY REMARKS

3.1. An Inherent Limitation of SAS

The property measured by the Y_a-statistics may, in the case of small areas, be expected to be subject to faster and larger temporal variations than holds true for large areas. Hence, if SAS are to be useful, estimates of the Y_a-statistics must be especially timely. This supports—as mentioned in Section 2.2—the reliance on some *indirect* estimation approach.

3.2. The Terminology

The specific SAS terminology is not entirely satisfactory. I will give two illustrations.

First, there appears to be no intrinsic difference between SAS and "domain statistics"; possibly, SAS may be viewed to be a special case of "domain statistics." Second, it has been argued that "synthetic estimation" is but a special case of "regression estimation". If so, do we need both terms?

3.3. Standards for SAS

Outside the realm of SAS, a variety of statistical standards have long been used as a guide to efficient survey design and implementation. Five examples of such standards are:

i. Standards for relevance and responsiveness.
ii. Standards for accuracy.
iii. Reproducibility standards.
iv. Minimum performance standards.
v. Standards of presentation.

Clearly, the need for standards for the production of SAS is no less than that for the production of other types of statistics. It is, however, not obvious that it would be realistic to aim at the same *high* standards for SAS as elsewhere.

I single out standards for accuracy for some additional comments. We may ask what is a meaningful *measure* of the accuracy of SAS. Some writers seem to advocate:

i. The mean-square errors for the *individual* area estimates.
ii. An average of these measures as an *overall* measure.

The use of the mean-square errors for the individual area estimates appears reasonable. The use of their average as an overall measure is, in my view, problematic on (at least) two grounds. First, it is difficult to assess what it signifies with respect to the usefulness of the estimates.

Second (and especially if it is the only measure presented), it may conceal the possible weakness of some of the individual area estimates.

An alternative would be as follows. It is based on the (estimated) relative root-mean-square error for an area:

$$\frac{\sqrt{\text{MSE } \tilde{Y}_a}}{\tilde{Y}_a} \times 100 = p_a$$

The overall measure would then be $\#(p_a > k)$, this is, the number of areas for which p_a exceeds a critical value k, possible chosen in cooperation with the users. If the number of such areas is large, it may be preferable to publish a frequency table of p_a-values.

4.   SOME POTENTIAL TOPICS FOR RESEARCH AND DEVELOPMENT

4.1.   The Future of SAS

SAS is here to say. We may indeed expect that the demands for SAS will increase in the years ahead.

Whether this increase proves to be large or only moderate, it will be a challenge to the statisticians to improve their performance in the realm of SAS. In Sections 4.2–4.5, I will present four topics for research and development.

4.2.   Using a Multivariate Estimation Algorithm

The algorithm $f(X_a)$ considered in Sections 2.3–2.4 is univariate: it involves only the X_a-statistics. It may be of interest to consider a multivariate algorithm, say $f(X_a, Z_a, \ldots, U_a)$.

4.3.   Creating a Taxonomy of Applications

A review of applications of various techniques $T_1 \cdots T_g$ for producing SAS suggests that no single technique is uniformly best for all kinds of situations $S_1 \cdots S_n$. Thus, T_1 may outperform the other techniques when applied in situation S_1, but being outperformed by T_2 in some other situation.

It may be worthwhile to subject this casual observation to an analysis of what has caused it: Is it just an accidental occurrence or does it reflect some mechanism that depends on the specific situation in which the technique is applied? If it appears reasonable to attribute the observation to such a mechanism, it may be worthwhile to develop a taxonomy of applications that would then serve as a guide to the choice of technique.

4.4. Exploiting Methods from Related Fields

If we agree that the estimation to produce SAS is in principle equivalent with (or at least very similar to) estimation of domain statistics, it may prove useful to assess methods from the literature on domain statistics from the point of view of their applicability in the realm of SAS.

The idea of "borrowing methods" may be taken one step further. Some fields to which we may turn in order to look for useful technical ideas are, to give but three examples:

(ii) Imputation.
(ii) Pooling estimates.
(iii) Smoothing.

4.5. Simulation

I will give a very simple example of how simulation may prove helpful to understand the properties of an estimation technique.

Consider a situation where some linear function $f(\cdot)$ serves well to model the data-generating system. The question is now: How strong must the correlation between the Y_a- and X_a-statistics be to make $f(\cdot)$ produce "accurate" estimates? To get an answer to this question, one set of X_a-statistics is chosen, and three sets of Y_a-statistics corresponding to three different correlations, say, corr = 0.20, 0.50, and 0.80. For each set of Y_a-statistics, the sets

$$Y^*_{i1} \cdots Y^*_{ia} \cdots Y^*_{iA}$$

$(i = 1, 2, 3)$ of estimates are computed and compared with the true values, using some chosen measure(s) of accuracy.

Discussion

G. Kalton
University of Michigan

As Brackstone observes in his paper at this symposium, there has been a marked increase in interest in small area data in recent years. Sample surveys nowadays produce rich data bases for a wide range of subjects, and they can provide sample-based estimates of adequate precision for numerous parameters at the national and major domain levels. However, even large-scale sample surveys are not able to provide sample-based estimates of adequate precision for small geographical areas; the sample sizes in small areas are insufficient to give useful sample-based estimates, and indeed a number of the areas may not be represented in the sample at all. Yet there is a considerable and growing demand for small area estimates from a variety of users: for instance, central governments increasingly need small area data for the allocation of program funds, local governments need such data for planning purposes, and market researchers need the data for determining effective marketing policies. It is this demand which cannot be satisfied by conventional sample survey estimation procedures that has led to the development of the wide range of small area estimation techniques that are the main focus of this symposium.

The essence of all small area estimation techniques is the use of model-dependent estimators. Many of the techniques employ completely model-dependent estimators, while others combine model-dependent and sample-based estimators [e.g., Fay and Herriot (1979)]. As an example of a model-dependent estimator, consider the estimation of the unemployment rates for small areas, using data from a large-scale unemployment survey. The simplest model is one that assumes that the small areas are microcosms of the total population, in which case each of the small area estimates would be the overall sample estimate. This model is clearly

264

oversimplistic, and hence more realistic models are sought. Such models require the availability of auxiliary data for the small areas. Thus, for example, if the age/sex distributions of the populations in the small areas are known, they may be used together with the sample age/sex-specific unemployment rates to compute the synthetic estimator of the unemployment rate for small area q as $p_q = \Sigma W_{hq} p_h$, where W_{hq} is the proportion of the population in small area q in the age/sex subgroup h, and p_h is the unemployment rate in the total sample for that subgroup. If there is regional variation in unemployment rates, the model may be modified to relate only to the region in which the small area lies, that is, p_h is taken to be the regional estimate of the unemployment rate for subgroup h.

The effect of incorporating the auxiliary variables into the synthetic estimator depends on two factors, the difference between the distribution of the small area and that of the total population across the subgroups created from the auxiliary variables, and the difference in the level of the dependent variable across the subgroups. If the age/sex distribution of the small area is the same as that of the total population, the use of the age/sex combination as an auxiliary variable in the synthetic estimator will yield the same estimate as that obtained from the simple model that uses no auxiliary variables. Similarly, if the unemployment rate does not vary by age/sex subgroup, the synthetic estimator will be the same as that obtained from the simple model. As with all model-based small area estimators, the quality of the synthetic estimator depends on the availability and proper use of appropriate auxiliary variables. In practice, the choice of auxiliary variables to be used is frequently severely restricted by the data available for the small areas. In many cases only a few auxiliary variables are available, and they are often only weakly related to the dependent variable. In such cases, the estimates for all the small areas will be close to one another, and hence they may not reflect the true variability in the population parameters. The quality of such estimates must be highly suspect. Another issue to be considered is whether the definitions of the auxiliary variables employed by the source of the small area statistics are the same as those used for the survey (see the chapter by Dagum et al. in this volume).

The development of good models for small area estimators requires in the first place the availability of the relevant auxiliary variables. Even if these variables are available, their application in an appropriate form is not straightforward. Evaluation studies of small area estimates have often shown up the presence of a number of outliers, which may be due to model misspecification. Thus, for instance, with U.S. states as the small areas, Schaible et al. (1979) found that synthetic estimators for the District of Columbia and Hawaii consistently performed badly, and in estimating union membership Heeringa (1981) found that the North and

South Carolina synthetic estimates were seriously in error. In this volume, Feeney identified a number of outliers in the small area estimates of unemployment for Local Government Areas in New South Wales. Sometimes, modifications of the model can be made to deal with the outliers, perhaps—as in Feeney's case—by restricting the application of the model to a region of the country. However, without such evaluation studies the outliers would go undetected, and some estimates would be seriously in error. Moreover, measures like the average mean square error of synthetic estimators over several small areas proposed by Gonzalez and Waksberg (1973) are unsatisfactory as indicators of the quality of individual estimates when outliers are present.

In view of the above issues, I consider that a cautious approach should be adopted to the use of small area estimates, and especially to their publication by government statistical agencies. When government statistical agencies do produce model-dependent small area estimates, they need to distinguish them clearly from conventional sample-based estimates. Some small area estimates may be seriously in error and, as Brackstone notes elsewhere in this volume, errors in small area estimates may be more apparent to users than errors in national estimates. Before small area estimates can be considered fully credible, carefully conducted evaluation studies are needed to check on the adequacy of the model being used. Sometimes model-dependent small area estimators turn out to be of superior quality to sample-based estimators, and this may make them seem attractive. However, the proper criterion for assessing their quality is whether they are sufficiently accurate for the purposes for which they are to be used. In many cases, even though they are better than sample-based estimators, they are subject to too high a level of error to make them acceptable as the basis for policy decisions.

REFERENCES

Fay, R. E., and Herriot, R. A. (1979), "Estimates of Income for Small Places: An Application of James-Stein Procedures to Census Data," *Journal of the American Statistical Association*, **74**, 269–277.

Gonzalez, M. E., and Waksberg, J. (1973), "Estimation of the Error of Synthetic Estimates," paper presented at the first meeting of the International Association of Survey Statisticians, Vienna, Austria, 18–25 August.

Heeringa, S. G. (1981), "Small Area Estimation Prospects for the Survey of Income and Program Participation," *Proceedings of the Section on Survey Research Methods*, American Statistical Association, pp. 133–138.

Schaible, W. L., Brock, D. B., Casady, R. J., and Schnack, G. A. (1979), *Small Area Estimation: An Empirical Comparison of Conventional and Synthetic Estimators for States*, Vital and Health Statistics, Series 2, No. 82. National Center for Health Statistics, Department of Health, Education, and Welfare, Hyattsville, MD.

Discussion

L. Kish
The University of Michigan

The past two days convinced me that a few words about classification and terminology may help to reduce confusion in this discussion and in the future. First, postcensal estimates for local areas belong clearly to a wider area of estimates for small domains (Purcell and Kish, 1979, 1980). For example, unemployment estimates for young Black males in the United States are even more vital than for an area of similar size (about 1%). Then let *domains* denote subpopulations, and let *subclasses* refer to their reflected subdivisions in the sample (see Table 1). *Strata* (though used for domains by sociologists) denote arbitrary subpopulations in sample designs, related to domains, but are usually more numerous. No clear limits for "small" are possible ("like beauty, small is in the eye of the beholder" said Dr. Wilk); but some tentative limits are needed to reduce confusion, and I was brave enough to try for some (Kish, 1980; Kish, 1987, Section 2.3).

For *major domains* reasonable sample-based estimates can be produced with standard methods from probability samples. These may be 20% or 10% or even 1% of the sample, depending on sample size and design, and on the precisions required. *Minor domains* comprise, say, less than 0.1 or even 0.01 of the population, hence separate estimates may be refused, but these days they are increasingly computed with "small domain estimates." *Mini domains* range from 0.01 to 0.001 or even to 0.0001, and censuses have been the traditional sources, but nowadays small domain estimation methods are tried for these also. *Rare items*, comprising less than 0.0001, pose problems for which overall samples are useless, and separate lists and methods are needed (Kish, 1965, Section 11.4).

In complex surveys (not srs) the effects, especially on variances,

267

Table 1. Classification of Domains and Subclasses (with Examples)

Types of Classes	Sizes of Classes					
	Major	0.1	Minor	0.01	Mini	0.0001
Design classes	Major regions. provinces		50 U.S. states		3000 U.S. counties	
Mixed	Partial segregation Mixed types		Natural resources Regions × age			
Crossclasses	5-year age groups Major occupations		Single years of age Occupation × education		Years age × education Age × education × income	

depend also on the *type* of the classes (Kish, 1980). *Design classes* denote domains (and subclasses) for which separate samples have been planned, designed, and selected; for example, major regions and urban and rural domains in most area samples; also geographical or alphabetical portions from appropriately ordered lists. At the other extreme are *cross-classes* that cut across the sample design, across strata, and across sampling units. These are the most commonly used domains and subclasses; for example, age, sex, occupation, education and income classes, behavior and attitude types, etc. They are not separated into design domains for lack of information or interest. Between the two extremes, and less common fortunately, are *mixed domains*, which are partially but not entirely segregated by natural or social factors; for example, fishermen, farming specialties, miners, and lumberjacks, but also some ethnic groups like "Blacks" and "Chicanos" in the United States.

1. AIMS OF DOMAIN ESTIMATES

Total population counts for small areas have dominated this field, partly because much research came from national census and statistical offices, also partly because of administrative and legal demands for better and current postcensal estimates. However, *other totals, means, and rates* are also obvious candidates and have been estimated; for example statistics for unemployment and housing (Gonzalez and Hoza, 1978), mortality, health, and disability (NCHS, 1968, 1977), employment (Laake, 1978), etc. Thus we need not only better postcensal estimates for local areas that are ordinarily collected by censuses, but also small domain estimates for variables not collected by censuses, nor by registers. Furthermore, *all statistics* of all kinds are candidates for separate domain estimates and

may benefit from separate consideration in multipurpose designs, I
believe. It has been only a simple, narrow view for statistical writing and
teaching to focus all design and analysis on the entire sample—whereas so
many statistical results are commonly presented for domains (Kish, 1961,
1969, 1980).

2. WAYS AND MEANS FOR DOMAIN ESTIMATES

Good *data sources* are the principal means to better statistics, many of us
believe. Furthermore, we may need to do more than merely discover and
utilize existing sources, but also to work toward the collection of other
and better auxiliary data; and that involves participation in public policy.
Second comes the methodological, mathematical, and theoretical
work of developing *better methods, formulas, designs, estimates, computations.* Third, we should consider, though they have been neglected here,
strategies for *cumulating data* from samples for small areas (NCHS, 1958;
Kish, 1981; 1987, Section 6.6).

3. METHODS FOR SMALL DOMAINS

Models, as well as data, are involved in all the diverse methods for small
domain estimates; they are involved in the choice of variables, in the
structural expressions for their use, and in the parameters for them. The
methods need only brief recall here (Purcell and Kish, 1979). *Symptomatic accounting techniques* have been developed by demographers, mostly
in the UN Census Bureau since before 1950 (Shryock and Siegel, 1975).
Regression-symptomatic methods are also old (Schmidt and Crosetti,
1954; Snow, 1911), but *synthetic* (ratio) estimates are newer (NCHS
1968).

The *sample regression* method (Ericksen, 1973) combines sample
survey data with symptomatic and census data. *Bayesian*, empirical
Bayes, and Stein–James methods propose to compromise (combine,
"shrink") local and global estimates (Efron and Morris, 1973). *SPREE* is
a "raking" method of iterative proportional fitting (IPF) and the most
general and flexible (Purcell and Kish, 1980).

Much of the above concentrates on local areas or on design domains.
Another strain of development dealt with cross-classes (Yates 1953,
Section 9.2; Kish 1961; Kish 1965, Section 4.5).

4. WHAT RESULTS?

We may bravely sketch the bare outlines of present and future developments. (1) Methods are *useful* and *used now* for postcensal estimates for local area statistics. The statistical offices of the United States, Canada, and Australia have been particularly prominent, but other offices are also active. (2) These methods *will be used for other statistics* also and in other situations. (3) The present methods can and *will be improved*. (4) The relative success of different methods is difficult to predict and it *depends on specific* circumstances; they must be discovered with specific empirical trials. (5) Success depends first on using better data and second on better methods.

REFERENCES

Efron, B., and Morris, C. (1973), "Stein's Estimation Rule and its Competitors—An Empirical Bayes Approach," *Journal of the American Statistical Association*, **68**, 117–130.

Ericksen, E. (1973), "A Method for Combining Sample Survey Data and Symptomatic Indicators to Obtain Population Estimates for Local Areas," *Demography*, **10**, 137–160.

Gonzalez, M. E., and Hoza, C. (1978), "Small Area Estimation with Application to Unemployment and Housing Estimates," *Journal of the American Statistical Association*, **73**, 7–15.

Kish, L. (1961), "Efficient Allocation of a Multipurpose Sample," *Econometrica*, **29**, 363–385.

Kish, L. (1965), *Survey Sampling*, Wiley, New York.

Kish, L. (1969), "Design and Estimation for Subclasses, Comparisons and Analytical Statistics," in *New Developments in Survey Sampling* (N. L. Johnson and H. Smith, eds.) Wiley, New York.

Kish, L. (1980), "Design and Estimation for Domains," *The Statistician* (Institute of Statisticians, London), **29**, 209–222.

Kish, L. (1987), *Statistical Design for Research*, Wiley, New York.

Laake, P. (1978), "An Evaluation of Synthetic Estimates of Unemployment,"*Scandinavian Journal of Statistics*, **5**, 57–60.

National Center for Health Statistics (1968), *Synthetic State Estimates of Disability*, PHS Publication 1759, U.S. Government Printing Office, Washington, D.C.

National Center for Health Statistics (1977), *Synthetic Estimation of State Health Characteristics*, PHS Publication 78-1349, U.S. Government Printing Office, Washington, DC.

Purcell, N. P., and Kish, L. (1979), "Estimation for Small Domains," *Biometrics*, **35**, 365–384.

Purcell, N. P., and Kish, L. (1980), "Postcensal Estimates for Local Areas (Small Domains)," *International Statistical Review*, **48**, 3–18.

Schmitt, R. C., and Crosetti, A. H. (1954), "Accuracy of the Ratio-Correlation Method for Estimating Population," *Land Economics*, **30**, 279-280.

Shryock, H. S., and Siegel, J. S. (1975), *The Methods and Materials of Demography*, U.S. Government Printing Office, Washington, DC.

Snow, E.C. (1911), "The Application of the Method of Multiple Correlation to the Estimation of Postcensal Populations," *Journal of the Royal Statistical Society*, **74**, 575–620.

Yates, F. (1953, 1960). *Sampling Methods for Censuses and Surveys*, Chas Griffin and Co., London.

Author Index

Almond, M. M., 59
Amemiya, Y., 106, 121

Bailar, B. A., 236
Balakrishnan, T. R., 59
Baldrige, 30, 44
Basavarajappa, K. G., 46, 49, 50, 61
Battese, G. E., 104, 122
Belsley, D. A., 63, 74
Bender, R. K., 46, 49, 50, 59, 61
Berry, B. J. L., 10, 20
Bettin, P. J., 27, 29, 44, 221, 223, 237
Blanchard, M. M., 104, 122
Blaustein, S., 173
Box, G. E. P., 128, 136
Brackstone, G. J., 3
Britton, M., 60
Brook, D. B., 248, 252, 266
Bryce, H. J., 236
Burtless, G., 166, 173

Cardenas, M., 104, 122, 123
Carr, T., 174
Carter, G., 24, 44
Casady, R. J., 266
Casella, G., 132, 136
Cassel, C. M., 141, 144, 159
Cavanaugh, F. J., 47, 61
Chandrasekar, C., 236
Chhikara, R. S., 104, 122
Choi, C. Y., 47, 59
Choudhry, G. H., 16, 20, 248, 252
Cliff, A. D., 59
Coale, A. J., 236
Cohen, S. B., 233, 236

Cook, D. R., 170, 174
Cook, P., 122, 123
Cowan, C. D., 27, 29, 44, 221, 223, 237
Cox, L. H., 17, 20
Craig, M. E., 104, 122, 123
Cronkhite, F. R., 160
Crosetti, A. H., 49, 63, 74
Cuomo, 30, 44
Czajka, J., 174

Dagum, E. B., 90, 123, 175, 176, 197, 265
Dalenius, T., 257
Deming, W. E., 236
Dempster, A. P., 77, 89, 90, 104, 122, 132, 137
Dielman, T. E., 169, 174
Diffendal, G. J., 219, 233, 237
DiGaetano, R., 104, 122
Draper, N. R., 163, 172, 174
Drew, J. D., 16, 20, 248, 252
DuMouchel, W., 24, 44

Effron, B., 92, 101, 104, 113, 114, 122, 269, 270
Ericksen, E. P., 16, 20, 23, 24, 25, 32, 38, 44, 45, 67, 74, 91, 101, 104, 122, 174, 227, 230, 231, 232, 233, 237, 269–270
Euler, 70

Fay, R. E., 24, 86, 90, 91, 92, 94, 95, 101, 104, 122, 124, 125, 132, 137, 145, 237, 264, 266
Feeney, G. A., 198, 266
Freedman, D. A., 24, 37, 38, 42, 43, 45

Fuller, W. A., 93, 101, 102, 103, 104, 122, 123
Furnival, G. M., 229, 237

Geisser, S., 102, 172, 174
Golberg, D., 49, 59
Goldberger, A. S., 169, 174
Gonzales, M. E., 104, 122, 142, 159, 177, 197, 252, 266, 268, 270
Goodnight, J. H., 89, 90
Grier, 49
Griffiths, W. E., 174

Hanuschak, G., 104, 122, 123
Harris, J., 24, 44
Harter, R. M., 93, 103, 122
Hartley, H. O., 93, 101
Harvey, B., 217
Harville, D. A., 93, 100, 102, 104, 106, 113, 114, 117, 122, 123
Hauser, P. M., 64, 74
Heeringa, S. G., 266
Henderson, C. R., 104, 123
Herman, 71
Herriot, R. A., 24, 87, 90, 92, 94, 95, 101, 104, 122, 124, 125, 132, 137, 237, 245, 264, 266
Hickman, R. D., 101, 102, 123
Hidoroglou, M. A., 78, 86, 90, 101, 102, 123, 175, 176, 179, 197
Hill, R. C., 174, 237
Hoza, C., 104, 122, 268, 270
Hughes, P. J., 47, 59

Isaki, C. T., 219, 233, 237

James, W., 25, 45, 92, 102, 104, 123
Johnson, T., 59
Jones, C. D., 237
Judge, G. G., 163, 166, 174

Kackar, R. N., 104, 117, 123
Kadane, J. B., 23, 24, 25, 32, 44, 45, 104, 122, 227, 230, 231, 232, 233, 237
Kalsbeek, W., 233, 236
Kalton, G., 264
Katz, M. B., 4, 20
Katzoff, M. J., 60
Kish, L., 16, 20, 63, 67, 74, 92, 102, 103, 123, 167, 174, 218, 237, 253, 267–270
Kitagawa, E. M., 63, 64, 74

Kleweno, D., 122
Kristiansson, K. E., 141
Kronecker, 89, 119
Kuh, E., 63, 74
Kullback, 65

Laake, P., 268, 270
Laird, N. M., 132, 137
Lee, T., 25, 133, 163, 166, 174
Lindley, D. V., 25, 45, 133, 137, 230, 237
Luebbe, R., 122
Lundstrom, S., 239, 240, 252
Leibler, 65

McCullagh, P., 62
Mackenzie, E., 122
Malec, D. J., 60, 233, 237
Mandell, M., 46, 49, 50, 60
Martin, J., 50, 60
Miller, A. J., 229, 237
Miller, C., 122
Morris, C. N., 24, 45, 92, 100, 102, 104, 113, 114, 122, 123, 132, 137, 269, 270
Morrison, P. A., 63, 64, 70, 71, 74
Morry, M. A., 90, 123, 175, 197
Musgrave, J., 171, 174

Nagi, James, 132, 137
Namboodiri, N. K., 63, 74
Namboodiri, N. R., 49, 59
Nargundkar, M. S., 17, 20
Nash, J. F., Jr., 64, 65, 72, 74
Navidi, W. C., 24, 37, 38, 42, 43, 45
Nebebe, Fassil, 132, 137
Norris, D. A., 60

O'Hare, N., 63, 71, 74
O'Hare, W., 49, 50, 51, 60
Ord, J. K., 59, 60
Ozga, M., 122

Parent, P., 60
Passel, J. S., 30, 145, 224, 234, 237, 238
Peixoto, J. L., 104, 113, 114, 123
Pickard, R. R., 170, 174
Pindyck, R. S., 166, 169, 174
Preston, S. H., 226, 236, 237
Purcell, N. J., 16, 20, 63, 67, 74, 92, 102, 103, 123, 167, 174, 199, 218, 237, 243, 253, 269

Raback, G., 141, 144, 159, 177, 197
Raby, R., 60
Raghunathan, T. E., 77
Rao, J. N. K., 90, 93, 101, 123, 197
Rao, V. R., 49, 59
Reinsel, G. C., 104, 123
Relles, D. A., 63, 64, 70, 71, 74
Rives, N. W., Jr., 236
Robbins, H., 104, 123
Robinson, J. G., 30, 45, 220, 224, 237
Rolph, 24, 44
Romaniuc, A., 60
Rosenberg, H., 63, 74
Rubin, D. B., 104, 122, 132, 137
Rubinfeld, D. L., 166, 169, 174

Sarndal, C. E., 86, 90, 104, 123, 142, 143,
    144, 159, 177, 179, 197, 242, 253
Saveland, W., 17, 20
Schaible, W. L., 248, 249, 253, 265, 266
Schirm, A. L., 226, 237
Schmitt, R. C., 49, 63, 74, 271
Schnack, G. A., 248, 253, 266
Schultz, L. K., 219, 233, 237
Searle, S. R., 93, 102, 123
Shiskin, J., 171, 174
Shyrock, H. S., 269, 271
Siegel, J. S., 30, 45, 220, 237, 269, 271
Sigman, R. S., 104, 123
Sigman, R., 123
Singh, M. P., 16, 20, 59, 248, 252
Slater, C. M., 237
Smith, A. F. M., 25, 45, 133, 137, 230, 237
Smith, H., 163, 174
Smith, P. J., 219

Snee, R. D., 172, 174
Snow, E. C., 271
Spar, M., 50, 60
Standish, L. D., 60
Stein, C., 25, 45, 92, 102, 104, 122, 123
Strawderman, W. E., 127, 128, 137
Stroud, T. W. F., 124, 125, 132, 137
Swanson, D. A., 49, 60, 71, 74

Tayman, J., 46, 49, 59, 60
Tedrow, L. M., 60, 71, 74
Tiao, G. C., 128, 136
Tsutakawa, R. K., 104, 122, 131, 132, 137
Tukey, J. W., 44, 45, 233, 237

Verma, R. B. P., 46, 49, 50, 59, 60, 61

Wahlstrom, S., 141
Waksberg, J., 122, 266
Walker, G., 123
Warren, R., 30, 45, 224, 238
Welsch, R. E., 63, 74
Werner, M., 217
Wilk, M., 267
Wilson, R. W., Jr., 229, 237
Wonnacott, R. J., 167, 174
Wonnacott, T. H., 167, 174

Yaffe, R., 12
Yates, F., 269, 271
Young, A., 171, 174

Zelnik, M., 236
Zidek, J. V., 62, 63, 74
Zitter, M., 47, 61

Subject Index

Absolute error, 50
Adhoc areas, 9
Adjustment factors, 220, 225, 226, 233, 234
Adjustment methods, 225, 227, 228, 231, 232, 234
Administrative records, 4, 13, 16, 63
Allocation structure, 202, 205, 206, 211
Area correlation, 87
Association structure, 202, 204

Bayesian estimates, 79, 87
Bayesian model, 78, 87, 131, 227, 230, 231
Bias, 64, 107, 109, 115, 118, 141, 143, 144, 148, 154, 156, 159, 203

Census adjustment, 230
Census evaluation, 224
Chi-square distribution, 127
Common Bayesian methods, 79
Component methods, 46, 49, 52, 57, 58
Confidentiality protection, 10
Conversion files, 10, 56
Covariate, 77, 86, 88, 93, 97, 99, 101, 106, 107, 111, 114, 117, 119, 124, 125, 126, 146, 232

Data development, 7, 9, 13, 14, 17
Data sources, 14, 221, 223
Degree of freedom, 103, 106, 115, 120, 127, 130, 131, 134, 135
Demographic analysis, 28, 29, 219, 221, 224, 225
Direct estimation, 16, 232, 233
Duel system, 222, 234

Empirical Bayes, 86, 91, 95, 100, 104, 125
Error of closures, 17
Estimated variance, 24, 109, 110, 151

Handbook, 163, 173
Horvitz–Thomson (HT) estimator, 241, 246

Iterative proportional fitting (IPF), 243, 244

Least squares, 48, 55, 66, 73, 105, 115, 130, 231
Linear model, 64, 73, 93
Linear regression, 91, 93, 94, 98, 227
Log linear model, 200

Minimal sufficiency statistics, 10
Missing data strategies, 26
Mixed models, 81, 101
Model bias, 154, 159
Model error, 95, 96, 203, 234
Model variance, 95
Monte Carlo simulation, 239, 240, 245, 248, 251, 252

Nonresponse rates, 147
Nonsampling errors, 37, 88
Normal distribution, 110, 113

Outliers, 82, 86, 204, 211

Posterior distribution, 126, 128, 129
Posterior mean, 100, 124, 127, 131, 132
Posterior variance, 126, 131, 132

Post stratified estimator, 86, 177, 178, 180, 182, 183, 190, 191, 195, 242, 246, 248, 251
Prediction variance, 107
Prior distribution, 125, 126, 127
Probability of response, 223

Ratio correlation, 46, 47, 48, 49, 50, 51, 53, 57, 63, 71
Regression coefficient, 32, 47, 48, 50, 52, 55, 58
Regression estimates, 23, 37, 40, 41, 42, 49, 52, 58, 95, 96, 131
Regression nested estimates, 49, 53, 54, 57, 58
Residual, 30, 95, 98, 135, 145, 201

Sample dependent estimation, 16
Sample regression, 67, 168
Sampling variance, 37, 39, 40, 95, 98, 154
Small area, 3, 6, 10, 11, 13, 14, 17, 18, 19, 46, 48, 58, 62, 72, 78, 103, 104, 112, 120, 124, 142, 143, 202, 203, 221, 224, 226, 234
Standard error, 38, 40, 47, 94, 96, 154, 155, 156, 204, 206, 217, 239, 248
Statistical significances, 201
Structure preserving estimate, 239, 240, 243, 244

Swedish Population and Housing Census, 239
Symptomatic indicators, 47, 50, 55, 56
Symptomatic variables, 63, 64, 66, 68, 70, 73
SYNC (count-synthetic estimator), 177, 178, 180, 182, 183, 185, 190, 193
SYN/R (ratio-synthetic estimator), 177, 178, 179, 180, 181, 182, 183, 185, 190, 191, 192, 193, 194, 195
Synthetic estimation, 16, 141, 142, 143, 144, 148, 152, 159

Time series, 217
Trend cycle, 171

Undercount rates, 24, 29, 30, 31, 35, 36, 37

Variance, 24, 37, 38, 80, 81, 93, 94, 97, 98, 99, 103, 104, 105, 106, 107, 108, 109, 110, 119, 120, 121, 124, 125, 126, 127, 128, 129, 131, 141, 151, 154, 205, 232, 233, 236
Variance estimator, 110
Variance of the model, 93, 99, 121, 127, 135
Variance predictor, 109, 120
Vital rate method, 46, 49

Weighting average, 96, 127

Applied Probability and Statistics (Continued)

JUDGE, HILL, GRIFFITHS, LÜTKEPOHL and LEE • Introduction to the Theory and Practice of Econometrics

JUDGE, GRIFFITHS, HILL, LÜTKEPOHL and LEE • The Theory and Practice of Econometrics, *Second Edition*

KALBFLEISCH and PRENTICE • The Statistical Analysis of Failure Time Data

KISH • Statistical Design for Research

KISH • Survey Sampling

KUH, NEESE, and HOLLINGER • Structural Sensitivity in Econometric Models

KEENEY and RAIFFA • Decisions with Multiple Objectives

LAWLESS • Statistical Models and Methods for Lifetime Data

LEAMER • Specification Searches: Ad Hoc Inference with Nonexperimental Data

LEBART, MORINEAU, and WARWICK • Multivariate Descriptive Statistical Analysis: Correspondence Analysis and Related Techniques for Large Matrices

LINHART and ZUCCHINI • Model Selection

LITTLE and RUBIN • Statistical Analysis with Missing Data

McNEIL • Interactive Data Analysis

MAINDONALD • Statistical Computation

MANN, SCHAFER and SINGPURWALLA • Methods for Statistical Analysis of Reliability and Life Data

MARTZ and WALLER • Bayesian Reliability Analysis

MIKÉ and STANLEY • Statistics in Medical Research: Methods and Issues with Applications in Cancer Research

MILLER • Beyond ANOVA, Basics of Applied Statistics

MILLER • Survival Analysis

MILLER, EFRON, BROWN, and MOSES • Biostatistics Casebook

MONTGOMERY and PECK • Introduction to Linear Regression Analysis

NELSON • Applied Life Data Analysis

OSBORNE • Finite Algorithms in Optimization and Data Analysis

OTNES and ENOCHSON • Applied Time Series Analysis: Volume I, Basic Techniques

OTNES and ENOCHSON • Digital Time Series Analysis

PANKRATZ • Forecasting with Univariate Box-Jenkins Models: Concepts and Cases

PIELOU • Interpretation of Ecological Data: A Primer on Classification and Ordination

PLATEK, RAO, SARNDAL and SINGH • Small Area Statistics: An International Symposium

POLLOCK • The Algebra of Econometrics

PRENTER • Splines and Variational Methods

RAO and MITRA • Generalized Inverse of Matrices and Its Applications

RÉNYI • A Diary on Information Theory

RIPLEY • Spatial Statistics

RIPLEY • Stochastic Simulation

RUBIN • Multiple Imputation for Nonresponse in Surveys

RUBINSTEIN • Monte Carlo Optimization, Simulation, and Sensitivity of Queueing Networks

SCHUSS • Theory and Applications of Stochastic Differential Equations

SEAL • Survival Probabilities: The Goal of Risk Theory

SEARLE • Linear Models

SEARLE • Matrix Algebra Useful for Statistics

SPRINGER • The Algebra of Random Variables

STEUER • Multiple Criteria Optimization

STOYAN • Comparison Methods for Queues and Other Stochastic Models

TIJMS • Stochastic Modeling and Analysis: A Computational Approach

TITTERINGTON, SMITH, and MAKOV • Statistical Analysis of Finite Mixture Distributions

UPTON • The Analysis of Cross-Tabulated Data

UPTON and FINGLETON • Spatial Data Analysis by Example, Volume I: Point Pattern and Quantitative Data

(continued from front)